SUSTAINABLE ENGINEERING

SUSTAINABLE ENGINEERING

Concepts, Design, and Case Studies

DAVID T. ALLEN
DAVID R. SHONNARD

PRENTICE
HALL

Upper Saddle River, NJ • Boston • Indianapolis • San Francisco
New York • Toronto • Montreal • London • Munich • Paris • Madrid
Capetown • Sydney • Tokyo • Singapore • Mexico City

The publisher offers excellent discounts on this book when ordered in quantity for bulk purchases or special sales, which may include electronic versions and/or custom covers and content particular to your business, training goals, marketing focus, and branding interests. For more information, please contact:

U.S. Corporate and Government Sales
(800) 382-3419
corpsales@pearsontechgroup.com

For sales outside the United States please contact:

International Sales
international@pearson.com

Visit us on the Web: informit.com/ph

Library of Congress Cataloging-in-Publication Data

Allen, David T.
   Sustainable engineering : concepts, design, and case studies / David T.
Allen, David R. Shonnard.
       p.   cm.
   Includes bibliographical references and index.
   ISBN 0-13-275654-4 (pbk. : alk. paper) 1. Sustainable engineering. 2. Environmental engineering. 3. Sustainable development. 4. Environmental management. I. Shonnard, David. II. Title.
   TA170.A42 2012
   620.8 — dc23   2011041250

ISBN-13: 978-0-13-275654-9
ISBN-10:   0-13-275654-4

Text printed in the United States on recycled paper at RR Donnelley in Crawfordsville, Indiana.
First printing, December 2011

Publisher: Paul Boger
Acquisitions Editor: Bernard Goodwin
Managing Editor: John Fuller
Project Editor: Elizabeth Ryan
Copy Editor: Barbara Wood

Indexer: Jack Lewis
Proofreader: Carol Lallier
Publishing Coordinator: Michelle Housley
Cover Designer: Chuti Prasertsith
Compositor: Laserwords Private Limited

Contents

2 RISK AND LIFE-CYCLE FRAMEWORKS FOR SUSTAINABILITY   35

3 ENVIRONMENTAL LAW AND REGULATION   65

4 GREEN, SUSTAINABLE MATERIALS   91

5   DESIGN FOR SUSTAINABILITY: ECONOMIC, ENVIRONMENTAL, AND SOCIAL INDICATORS   117

6   CASE STUDIES   165

Preface

Growing populations and affluence around the globe have put increasing pressure on natural resources, including air and water, arable land, and raw materials. Concern over the ability of natural resources and environmental systems to support the needs and wants of global populations, now and in the future, is part of an emerging awareness of the concept of sustainability. Developing new technologies that address societal needs and wants, within the constraints imposed by natural resources and environmental systems, is one of the most important challenges of the 21st century. Engineers will play a central role in addressing that challenge.

Engineers can design products, processes, and technologies that are sustainable by integrating environmental, economic, and social factors in the evaluation of their designs. While this may seem simple in the abstract, the tools for converting sustainability concepts into the types of quantitative design approaches and performance metrics that can be applied in engineering design are just emerging. These emergent tools of designing for sustainability are the topic of this textbook.

This book begins, in Chapter 1, with a brief review of some of the details of the natural resource and environmental challenges that engineers will face in designing for sustainability. Then, in Chapters 2 and 3, analysis and legislative frameworks for addressing environmental issues and sustainability are presented. This introductory material sets the stage for common frameworks used by engineers to incorporate sustainability into their designs. This material on designing sustainable engineered systems begins in Chapter 4 with a description of the selection of green and sustainable materials, since virtually every engineering design involves some use of materials. Chapter 5 then describes a series of principles that engineers can use to make engineering designs more sustainable and tools that can be used to evaluate, and in some cases monetize, the benefits of more sustainable designs. Finally, Chapter 6 presents case studies that illustrate the principles outlined in the text.

This book grew out of the authors' experiences in teaching methods of Green Engineering to chemical engineering students (Allen and Shonnard, 2002) and out of an understanding of emerging practices for incorporating sustainability into engineering curricula (Murphy et al., 2009). This book is designed to be an introduction to the concepts involved in designing sustainable systems, suitable for students in all engineering disciplines. The text emphasizes the economic and environmental dimensions of sustainability, since it is in these domains that quantitative tools suitable for engineering design are emerging. The social and societal dimensions of sustainability are examined, but at present, there are few tools available for explicitly incorporating societal and social objectives into engineering designs, and it is these quantitative tools that are the main focus of this text.

The book was made possible through the support of the Chemical Engineering Branch of the Office of Pollution Prevention and Toxics of the U.S. Environmental Protection Agency, and the hope of the authors and our colleagues at the EPA is that it will help direct the creativity of engineers toward the design of ever more sustainable engineered systems.

—*David T. Allen, University of Texas at Austin*
—*David R. Shonnard, Michigan Technological University*

REFERENCES

Allen, D. T., and D. R. Shonnard. 2002. *Green Engineering: Environmentally Conscious Design of Chemical Processes.* Upper Saddle River, NJ: Prentice Hall.

Murphy, C., D. T. Allen, B. Allenby, J. Crittenden, C. Davidson, C. Hendrickson, and S. Matthews. 2009. "Sustainability in Engineering Education and Research at U.S. Universities." *Environmental Science & Technology* 43:5558–64.

Acknowledgments

For more than a decade, we have been working with colleagues at the U.S. Environmental Protection Agency and at universities, developing materials on Green and Sustainable Engineering. The collaborations have included textbook development, short course offerings, creation of educational modules, and benchmarking activities. This textbook is the latest product of that collaboration, so while it bears our names as authors, it would not have been possible without the support of our colleagues at the EPA and other collaborators who have worked with us on these initiatives. Individual chapters acknowledge collaborators who assisted in the development of specific materials on which the material in this text is based, but overall guidance for this decade-long effort came from Sharon Austin and Nhan Nguyen in the Office of Pollution Prevention and Toxics at the EPA and from Bernard Goodwin at Prentice Hall. We are grateful for their leadership and vision. We also thank our peers, who reviewed drafts of the manuscript, and our students, whose questions helped us refine and clarify our work.

—*David T. Allen*
—*David R. Shonnard*

About the Authors

Dr. David T. Allen is the Gertz Regents Professor of Chemical Engineering, and the director of the Center for Energy and Environmental Resources, at the University of Texas at Austin. He is the author of multiple books and hundreds of scientific papers in areas ranging from coal liquefaction and heavy oil chemistry to the chemistry of urban atmospheres. The quality of his work has been recognized by research awards from the National Science Foundation, the AT&T Foundation, the American Institute of Chemical Engineers, the Association of Environmental Engineering and Science Professors, and the State of Texas. The findings from his research have been used to guide air quality policy development, and he has served on the U.S. EPA's Science Advisory Board and the National Research Council's Board on Environmental Studies and Toxicology, addressing issues at the interface between science, engineering, and public policy. For the past two decades, his work has also focused on the development of materials for environmental education, including coauthoring the textbook *Green Engineering: Environmentally Conscious Design of Chemical Processes*. He has won teaching awards at the University of Texas and UCLA. Dr. Allen received his B.S. in chemical engineering, with distinction, from Cornell University in 1979. His M.S. and Ph.D. degrees in chemical engineering were awarded by the California Institute of Technology in 1981 and 1983. He has held visiting faculty appointments at the California Institute of Technology, the University of California, Santa Barbara, and the Department of Energy.

Dr. David R. Shonnard is Robbins Professor in the Department of Chemical Engineering at Michigan Technological University and director of the Sustainable Futures Institute. He received a B.S. in chemical/metallurgical engineering from the

University of Nevada, Reno, in 1983; an M.S. in chemical engineering from the University of California, Davis, in 1985; a Ph.D. from the University of California, Davis, in 1991; postdoctoral training in bioengineering at the Lawrence Livermore National Laboratory from 1990 to 1993; and he was a visiting instructor at the University of California at Berkeley in 2003. His experiences in life-cycle assessment (LCA) methods and applications include a one-year sabbatical at the Eco-efficiency Analysis Group at BASF AG in Ludwigshafen, Germany. He has been on the faculty in the Department of Chemical Engineering at Michigan Technological University since 1993. Dr. Shonnard has more than twenty years of academic experience in sustainability issues in the chemical industry and Green Engineering. He is coauthor of the textbook *Green Engineering: Environmentally Conscious Design of Chemical Processes,* published by Prentice Hall in 2002. His current research interests focus on investigations of new forest-based biorefinery processes for production of transportation fuels, such as cellulosic ethanol and pyrolysis-based biofuels, from woody biomass using recombinant DNA and other approaches. Another active research area is LCA of biofuels and other biorefinery products to determine greenhouse gas emissions and net energy balances. He has contributed to National Academy of Sciences publications on green chemistry/engineering/sustainability in the chemical industry. Dr. Shonnard has coauthored 70 peer-reviewed publications and received numerous honors and awards for teaching and research into environmental issues of the chemical industry, including the Ray W. Fahien Award from ASEE (2003). He is a recipient of the NSF/Lucent Technologies Foundation Industrial Ecology Research Fellowship (1998) for research that integrates environmental impact assessment with process design.

CHAPTER 1

An Introduction to Sustainability

1.1 INTRODUCTION

Environmental and natural resource issues have gained increasing prominence in the latter half of the 20th century and the beginning of the 21st century. Growing populations and affluence, around the globe, have put increasing pressure on air and water, arable land, and raw materials. Concern over the ability of natural resources and environmental systems to support the needs and wants of global populations, now and in the future, is part of an emerging awareness of the concept of sustainability.

Sustainability is a powerful, yet abstract, concept. The most commonly employed definition of sustainability is that of the Brundtland Commission report—meeting the needs of the present generation without compromising the ability of future generations to meet their needs (World Commission on Environment and Development, 1987). However, a search on the definition of sustainability will return many variations on this basic concept. In engineering, incorporating sustainability into products, processes, technology systems, and services generally means integrating environmental, economic, and social factors in the evaluation of designs. While the concepts of engineering for sustainability may seem simple in the abstract, converting the concepts into the quantitative design tools and performance metrics that can be applied in engineering design is a challenge. Addressing that challenge is the topic of this textbook.

Quantitative tools available to engineers seeking to design for sustainability are continually evolving, but currently they focus on natural resource conservation and emission reduction. Few quantitative tools are currently available for incorporating social dimensions of sustainability into engineering design tools, and consequently, these issues receive limited treatment in this text.

Before beginning an examination of engineering concepts and tools, it is useful to first review some of the details of the natural resource and environmental challenges that engineers face in designing for sustainability. These topics are covered in the sections that follow.

1.2   THE MAGNITUDE OF THE SUSTAINABILITY CHALLENGE

To grasp the magnitude of the pressures on resources and ecosystems, it is useful to invoke a conceptual equation that is generally attributed to Ehrlich and Holdren (1971). The equation relates impact (I) to population (P), affluence (A), and technology (T):

$$I = P*A*T$$

This conceptual relationship, referred to as the *IPAT equation*, suggests that impacts, which could be energy use, materials use, or emissions, are the product of the population (number of people), the affluence of the population (generally expressed as gross domestic product, GDP, of a nation or region, divided by the number of people in the nation or region), and the impacts associated with the technologies used in the delivery of the affluence (impact per unit of GDP). For example, if the IPAT equation were used to describe energy use in the United States, I would represent energy use per year, P would represent the population of the United States, A would represent the annual GDP per capita, and T would represent the energy use per dollar of GDP.

While the IPAT equation should not be viewed as a mathematical identity, it can be used to assess the magnitude of the challenges that our societies face in materials use, energy use, and environmental impacts. By estimating growth in population and affluence, we can get an indication of the amount by which energy use, materials use, and emissions might increase over the next several decades, if our technologies remain static. Estimates from the United Nations (United Nations, 2007) suggest that world population will increase at the rate of 1% to 2% per year until peaking at somewhere near 10 billion, over the next century. Affluence, as measured in economic output (e.g., GDP), is growing in some regions of the world by 8% to 10% per year. On average, worldwide, affluence is growing by roughly 2% to 4% per year, depending on economic conditions. If these trends continue for several decades, compounded growth would lead world economic output (P*A) to increase by 50% in 10 years, by 300% in 25 years, and by more than a factor of 10 in 50 years.

Invoking the IPAT equation, the implications of population and economic growth are that if technology remains static, energy use, materials use, and environmental impacts will grow 10-fold over the next 50 years. Reducing the impacts of technology (T in the IPAT equation) by an order of magnitude will be necessary if the world is to support 10 billion people, all aspiring to better living standards. Reducing energy use, materials use, and emissions will be a central challenge for engineers of the 21st century, and engineers will need to develop and master technical tools that will integrate the objectives of energy efficiency, materials efficiency, and reduced environmental emissions into design decisions.

1.3   ENERGY

Energy is required for all economic activity. Inexpensive energy makes possible a high standard of living and many of the conveniences enjoyed in modern societies, such as heating, lighting, electronic devices, travel, and virtually all forms of communications. Therefore, an understanding of global energy supplies, and the environmental impacts associated with energy production and use, is important in understanding our ability to sustain current standards of living.

Energy is often categorized as renewable or nonrenewable. Examples of renewable energy sources include solar radiation, wind, and biomass. Fossil fuels (crude oil, coal, and natural gas) are nonrenewable energy sources because of the long periods necessary for regeneration, which can be on the order of millions of years. As the 21st century begins, the world relies primarily on fossil energy. Petroleum, coal, and natural gas supply 86% of world energy supplies, as shown in Figure 1-1.

From 1980 until 2006, world energy use increased from just under 300 quadrillion (10^{15}) BTU (quads) to more than 450 quads (an increase of more than 50%). Currently, fossil fuels make up 86% of the world's energy consumption (EIA, 2009a), while renewable sources such as biomass (including animal waste), hydroelectric, solar, and wind power account for only about 8% of the energy supply. Nuclear energy provides roughly 6% of world energy demand.

Similar energy data for the United States are shown in Figure 1-2. The left-hand side of the figure shows that the distribution of energy sources in the United States is

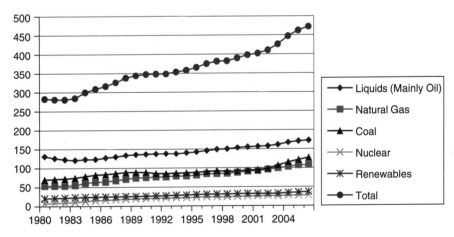

Figure 1-1   World energy use in quadrillion (10^{15}) BTU (quads) (EIA, 2009a)

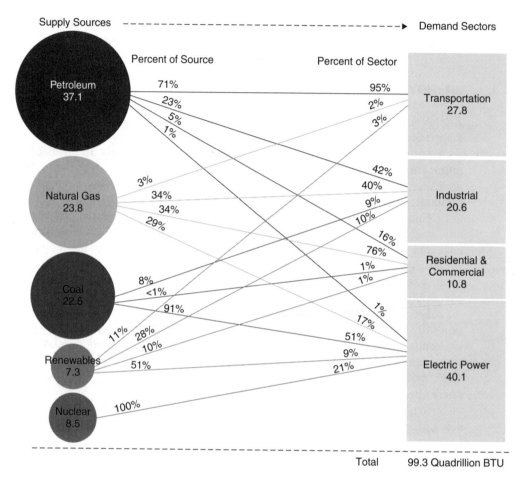

Figure 1-2 Energy sources and uses in the United States, expressed in quads per year; lines connecting energy sources and demand sectors show the percentage of energy sources provided to various demand sectors (left) and the percentage of the demand provided by various sources of energy (right). (EIA, 2009b)

similar to that of worldwide energy sources. Fossil fuels account for 84% of energy use; nuclear and renewables account for roughly equal portions of the remainder. The only significant difference between the U.S. patterns of energy supply and worldwide patterns is the source of biomass-derived fuels. In the United States, ethanol derived from corn was the dominant biomass-derived fuel, whereas worldwide, animal waste used as fuel and ethanol derived from sugarcane were the dominant biomass-derived fuels.

The right-hand side of Figure 1-2 shows the patterns of energy use in the United States, where more than 40% of the energy used is used to generate electricity. The remaining nonelectrical uses are transportation, non-electricity-generating industrial uses, and residential and commercial (largely building) energy use. Lines, with percentages noted at both ends, connect energy sources and energy uses in Figure 1-2. The percentages on the left-hand side of the diagram show the ways in which different fuels are used. The percentages on the right-hand side of the diagram show the types of fuels that are used for different energy applications. The percentages make clear that not all energy sources can be used in all applications. For example, 100% of nuclear power and 91% of coal are used to generate electricity. These fuels are not widely used for other purposes. Transportation relies almost exclusively (95%) on petroleum. Few other fuels are used for this application.

Example 1-1 Per capita energy use

Using the data in Figures 1-1 and 1-2, compare per capita annual energy use in the United States and worldwide. Assume a U.S. population of 300 million and a world population of 7 billion. Convert the result, expressed in BTU per person, into the number of gallons of gasoline that would be required to provide the energy. Assume that the energy content (higher heating value) of gasoline is 124,000 BTU/gal (Argonne National Laboratory, 2011).

Solution: Per capita energy consumption in the United States is more than five times greater than the worldwide average.

U.S. per capita annual energy use $= 99.3 * 10^{15}$ BTU/300 $* 10^6$ people $= 330 * 10^6$ BTU/person.
Global per capita annual energy use $= 450 * 10^{15}$ BTU/7 $* 10^9$ people $= 64 * 10^6$ BTU/person.
To convert BTU into gallons of gasoline equivalent, assume that the energy content of gasoline is 124,000 BTU/gal.
U.S. per capita annual energy use $= 330 * 10^6$ BTU $= 2700$ gallons of gasoline equivalent per year.
Global per capita annual energy use $= 64 * 10^6$ BTU $= 520$ gallons of gasoline equivalent per year.

Another way to map energy use is shown in Figure 1-3. This mapping, produced by Lawrence Livermore National Laboratory (2010), shows energy sources on the left and energy uses on the right, as in Figure 1-2. Figure 1-3 distinguishes between domestic and imported sources of energy, showing that the primary form of imported energy in the United States is petroleum. In mapping the flow from energy sources to energy uses, Figure 1-3 shows energy losses. These losses come about for a variety of reasons. The largest single category of loss is in electric power generation, amounting to more than a quarter of all U.S. energy use. This loss is in part a result of the laws of thermodynamics, since generating electrical power generally involves converting heat (from burning fuels) into mechanical work (turning a generator with steam made from the heat from burning the fuel). This "waste" heat is not really lost; energy is conserved. Instead, it is simply available at too low a temperature to be of significant economic value in power generation, so it is released into the environment as hot water or hot gases. Example 1-2 illustrates typical energy losses from fuel source to final end use.

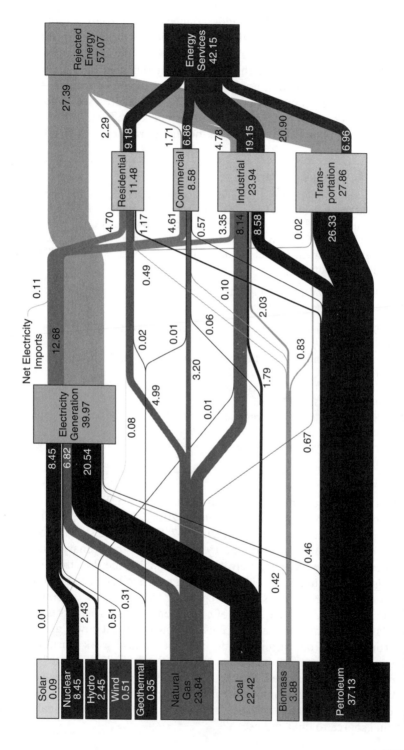

Figure 1-3 Mapping of energy sources and uses in the United States for 2008, quantifying energy losses (Lawrence Livermore National Laboratory, 2010)

Example 1-2   Efficiency of energy use

Determine the efficiency of energy utilization for a pump. Assume the following efficiencies in the energy conversion:

- Crude oil to fuel oil is 90% (0.90) (i.e., the energy required to produce and refine crude oil consumes 10% of the energy of the crude oil input to the process).
- Fuel oil to electricity is 40% (0.40) (i.e., the conversion of thermal energy into electrical energy occurs with an efficiency of 40%, roughly the average for the U.S. electrical grid).
- Electricity transmission and distribution is 90% (0.90) (i.e., losses of electricity in transmission from the power plant to the point of use are 10%).
- Conversion of electrical energy into mechanical energy of the fluid being pumped is 40% (i.e., the efficiency of the pump in converting electrical energy into the mechanical energy of the fluid is 40%).

Solution: The overall efficiency for the primary energy source is the product of all of the individual conversion efficiencies.

Overall efficiency = (0.90)(0.40)(0.90)(0.40) = 0.13 or 13%.

As shown in Figure 1-3, more than half of the energy used in the United States is lost in the process of converting the energy into useful forms, such as electricity. As shown in Example 1-2, the losses that occur between the source of the primary energy and the desired outcome of using the energy (such as moving a fluid through a pipe, or delivering horsepower at the wheel of a car, or illuminating a room with a lightbulb; see the problems at the end of the chapter) can be much larger. Improving energy efficiency is the work of engineers, and as shown in Figure 1-4, engineers have continuously improved the energy required per unit of economic output (the T in the IPAT equation) in both developed and developing economies.

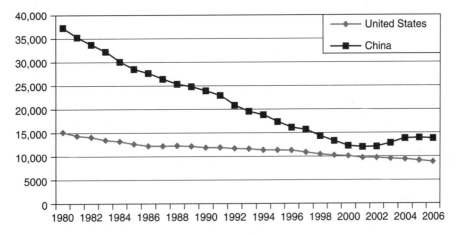

Figure 1-4 World energy intensity—total primary energy consumption per dollar of GDP using purchasing power parities (BTU per U.S. dollar (2000)) (EIA, 2009a)

Improving energy efficiency can reduce costs and conserve natural resources. It can also reduce environmental impacts. A variety of environmental impacts are associated with energy consumption. Fossil fuel combustion releases carbon dioxide into the atmosphere. Because of its long residence time in the atmosphere and because it readily absorbs infrared radiation, CO_2 can cause global warming. Furthermore, combustion releases oxides of nitrogen and, for some fuels, sulfur into the air, where photochemical reactions can convert them into ground-level ozone and acid rain. Hydropower energy generation causes land inundation, habitat destruction, and alteration in surface and groundwater flows. Nuclear power has environmental problems linked to uranium mining and spent nuclear fuel disposal. Renewable fuels are not environmentally benign either. Traditional energy usage (wood) has caused widespread deforestation in localized regions of developing countries. Solar power panels require energy-intensive use of heavy metals and creation of metal wastes. Satisfying future energy demands must occur with a full understanding of competing environmental, natural resource, and energy needs.

1.4   MATERIALS USE

Just as energy use is a required basis of economic activity, so is materials use; and just as energy use has increased as population and affluence have increased, so has materials use. This section will examine materials use in the United States and the world, without focusing on specific materials. The exception will be water. The amount of water used is so extensive, it would dominate any examination of materials use if it were not addressed separately.

1.4.1   Minerals, Metals, and Organics

Figures 1-5 and 1-6 show trends in world and U.S. materials use, over multiple decades.

Example 1-3   Per capita materials use

Using the data in Figure 1-6, compare per capita annual materials use in the United States and worldwide. Assume a U.S. population of 300 million and a world population of 7 billion. How much does the result change if the mass of fuel used is added? As a rough approximation of the mass of fuel used, use the per capita use of gasoline equivalents from Example 1-1 (e.g., 2700 gallons of gasoline equivalent per year for the United States) multiplied by the density of gasoline (about 6 lb/gal).   *Vol × Density*

Solution: Per capita materials consumption in the United States is more than six times greater than the worldwide average.

U.S. per capita annual materials use = $6 * 10^{12}$ lb/$300 * 10^6$ people = 20,000 lb/person per year.
Global per capita annual energy use = $21 * 10^{12}$ lb/$7 * 10^9$ people = 3000 lb/person per year.

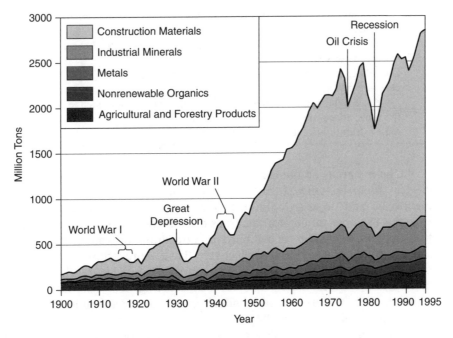

Figure 1-5 Materials use in the United States over the 20th century (USGS, 2002)

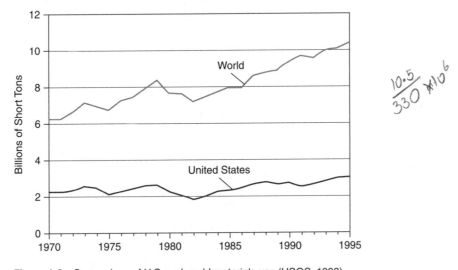

Figure 1-6 Comparison of U.S. and world materials use (USGS, 1998)

An estimate of the mass of fuel used can be made by multiplying per capita use of gasoline equivalents by the density of gasoline.

U.S. per capita annual materials use = 20,000 lb/person + 2700 gal * 6 lb/gal = 36,000 lb/person per year.

Global per capita annual materials use = 3000 lb/person + 520 gal * 6 lb/gal = 6100 lb/person per year.

If we express the mass of fuel use as the amount of carbon dioxide released by burning the fuels, accounting for oxygen use, rather than as the amount of fuel consumed, the mass attributed to fuel use more than triples.

The materials accounted for in Figures 1-5 and 1-6 do not include materials used as fuels. Including fuels doubles the materials use estimates. The data in Figures 1-5 and 1-6 also do not include material flows that do not enter the economy. For example, mining for materials often involves removing the ground above the ore seam, referred to as overburden. Agricultural operations involve the loss of soil into waterways as runoff. Collectively, these indirect flows are roughly as large as the total of all flows that enter the economies of highly developed nations (for more details, see Adriaanse et al., 1997). Considering all of these flows, a reasonable estimate of the materials used in industrialized countries is more than 100 pounds per person per day, not including water.

Many of these materials are limited natural resources. Consider metals as an example. If the total amount of materials available as ores, recoverable using current technologies, is divided by annual global consumption rates, only iron and aluminum, of the industrial metals, have economically recoverable reserves that would last more than 100 years (Graedel and Allenby, 1995). Further complicating resource availability is the issue of where the material resources are located. Many of the scarcest reserves are located in politically unstable regions of the world (USGS, 2002).

Increasing scarcity of materials and concerns about releasing waste materials into the environment will likely drive engineers to design systems that reuse and recycle materials. Lead (Pb) provides a case study of this evolution toward more recycling and reuse of materials. Figure 1-7 shows the increasing recycling and reuse of lead between 1970 and the mid-1990s. The figure is formatted in a manner analogous to Figure 1-3. Supply is documented on the left side of the diagram and uses are shown on the right. For materials, however, unlike energy, there are recycle loops. In the case of lead, the recycle loops are dominated by the flows of recycled batteries.

Example 1-4   Recycling rates

Using the data in Figure 1-7, compare the fraction of lead that was recycled in 1970 with that recycled in the mid-1990s.

Solution: Lead recycling doubled from approximately one-third of lead use to two-thirds of lead use from 1970 to the mid-1990s.

Fraction recycled in 1970 = 450,000 metric tons (mt)/1,230,000 mt = 0.36.
Fraction recycled in mid-1990s = 910,000 mt/1,410,000 mt = 0.65.

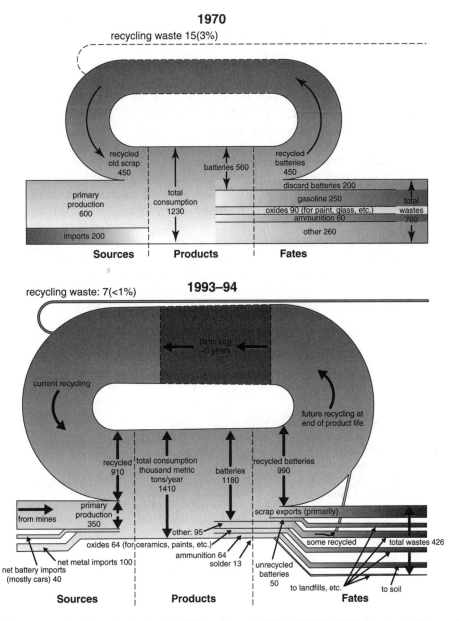

Figure 1-7 Material cycles for lead in 1970 and the mid-1990s (USGS, 2000); flows are in thousands of metric tons per year

Metals are not the only materials that can be reused and recycled. A number of case studies have been described by Allen (2004).

1.4.2 Water

Water is essential to life, and water use exceeds the use of any other substance. Although water is abundant, clean freshwater suitable for agriculture, industrial uses, or satisfying thirst is becoming increasingly scarce. As shown in Figure 1-8, freshwater use in the United States totals approximately 345,000 million gallons per day. This translates to approximately 1000 gallons per person per day.

The sources include surface waters, such as lakes and rivers, and groundwater. While surface water is renewed on relatively short timescales, groundwater, in some cases, is a resource that has accumulated over very long periods of time and may or may not be replaced at the same rate at which it is withdrawn.

The largest uses of water are for agriculture, thermoelectric power (power generation using a steam cycle), and public supplies. Although not shown in Figure 1-8, uses of water are generally categorized into withdrawals and consumption. Water withdrawals involve removing water from a surface or groundwater reservoir. If that water is not returned to the same reservoir, the water is referred to as having been

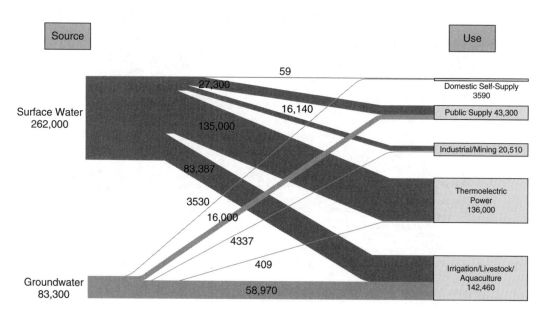

Figure 1-8 Water sources and uses in the United States, reported in millions of gallons per day. Numbers shown may not add to totals because of independent rounding. (Lawrence Livermore National Laboratory, 2004)[1]

1. In addition, 62,300 Mgal/day of saline water was withdrawn, primarily for thermoelectric use.

$E = 6T^4$

consumed. For example, if a power plant withdraws water from a lake, the total amount removed from the lake is the withdrawal. If the power plant uses some of that water to make steam to drive a turbine, and the steam is released to the atmosphere, the amount released to the atmosphere is considered consumption.

Example 1-5   Water use in electricity generation

Total electricity generation in the United States in 2008 was 12.68 quads (Figure 1-3). The water used (withdrawn) by the power sector was 136,000 Mgal/day. Calculate the amount of water used per kilowatt-hour of electricity generated.

Solution: Water used = 136,000 Mgal/day * 365 day/yr = 5 * 10^{13} gal/yr.
Electricity produced = 12.68 * 10^{15} BTU/yr * 1 kWh/3413 BTU = 3.7 * 10^{12} kWh.
Average water use = 13 gal/kWh.

$R_E (1-A) S = 4 \cdot E \cdot R_E$

This extensive use of water per unit of economic output (a kilowatt-hour retails for about 10 cents) is not unusual. Mining a million dollars' worth of coal requires 11 million gallons of water. Making a million dollars' worth of automobiles or semiconductors requires about 9 million gallons of water (EIOLCA, 2011). Clearly, our engineered systems are extensive users of water, and our economies depend on readily available, inexpensive, clean water.

BTU   Unit of work   1055 J   1HP = 746W   W → Power.

1.5   ENVIRONMENTAL EMISSIONS

Camelina derived jet fuel Ediesel: such odv biodiesel   kWh → energy unit

Environmental emissions and their impacts are global, regional, and local in scope. They also act on timescales ranging from hours to decades. While there are many types of environmental emissions, in this section atmospheric emissions and their impacts will be used to provide an example of the different spatial and temporal scales associated with environmental impacts. Readers seeking additional information on environmental emissions and their impacts can find more information in the U.S. Environmental Protection Agency's *Report on the Environment* (2008).

On a global scale, man-made (anthropogenic) greenhouse gases, such as chlorofluorocarbon (CFC) refrigerants, methane, nitrous oxide, and carbon dioxide, are implicated in global warming and climate change. These emissions have the potential to alter global weather patterns and may cause dislocation of people and animal species. On a regional scale, hydrocarbons released into the air, in combination with nitrogen oxides originating from combustion processes, can lead to air quality degradation over urban areas and over regions extending hundreds to thousands of kilometers. On a local scale, emissions of certain toxic air pollutants can lead to neighborhood-level concerns. Temporal scales of these atmospheric impacts also vary. For example, greenhouse gases typically have atmospheric lifetimes on the order of several decades. Other releases, such as those that impact urban air quality, can have their primary impact over a period of hours or days.

1.5.1 Ozone Depletion in the Stratosphere

There is a distinction between "good" and "bad" ozone (O_3) in the atmosphere. Tropospheric ozone, created by photochemical reactions between nitrogen oxides and hydrocarbons at the Earth's surface, is an important component of smog. A potent oxidant, it causes lung irritation that can lead to serious lung damage, and it damages crops and trees. Stratospheric ozone, found in the upper atmosphere, performs a vital and beneficial function for all life on Earth by absorbing harmful ultraviolet radiation. The potential destruction of this stratospheric ozone layer is therefore of concern and represents a global environmental challenge.

The stratospheric ozone layer is a region in the atmosphere between 6 and 30 miles (10–50 km) above ground level in which the ozone concentration is elevated compared to all other regions of the atmosphere. In this low-pressure region, the concentration of O_3 can be as high as 10 parts per million (10 ppm, 10 out of every 1,000,000 molecules). In contrast, ozone concentrations at ground level that exceed 100 parts per billion (100 ppb, 100 out of every 1,000,000,000 molecules) are considered a threat to human health. Ozone is formed at altitudes between 25 and 35 km in the tropical regions near the equator where solar radiation is consistently strong throughout the year. Because of atmospheric motion, ozone migrates to the polar regions, and its highest concentration is found there at about 15 km in altitude. Stratospheric ozone concentrations have declined over the past 30 years. The potential negative impact on human health and on a variety of plants has caused a considerable amount of concern among scientific and medical personnel worldwide. An excess of UV radiation can cause skin cancer in humans and damage to vegetation.

Ozone equilibrates in the stratosphere as a result of a series of natural formation and destruction reactions that are initiated by solar energy. The natural cycle of stratospheric ozone chemistry is being altered by the introduction of man-made chemicals. Three chemists, Mario Molina and Sherwood Rowland of the University of California, Irvine, and Paul Crutzen of Germany, received the 1995 Nobel Prize for chemistry for their discovery that CFCs take part in the destruction of atmospheric ozone. CFCs are highly stable chemical structures composed of carbon, chlorine, and fluorine. One important example is trichlorofluoromethane, CCl_3F, or CFC-11.

CFCs reach the stratosphere because of their chemical properties: high volatility, low water solubility, and persistence in the lower atmosphere. In the stratosphere, they are photo-dissociated to produce chlorine atoms, which then catalyze the destruction of ozone (Molina and Rowland, 1974):

$$Cl + O_3 \rightarrow ClO + O_2$$
$$ClO + O \rightarrow O_2 + Cl$$
$$\overline{O_3 + O \rightarrow O_2 + O_2}$$

The chlorine atom is not destroyed in the reaction cycle and can cause the destruction of thousands of molecules of ozone before forming HCl by reacting with hydrocarbons. The HCl eventually precipitates from the atmosphere. A similar

Sustainable Development Progress metrics

Institute of Chem Engr.

Triple Bottom Line env. responsibility, wealth creation,
social development.

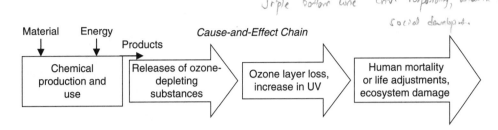

Figure 1-9 Ozone-depleting chemical emissions and the major steps in the environmental cause-and-effect chain

green Engg Educ in Chemical Engg Curricula ...
A Quarter Century of Progress for future Generations

mechanism also applies to bromine, except that bromine is an even more potent ozone-destroying compound. Figure 1-9 summarizes the major steps in the environmental cause-and-effect chain for ozone-depleting substances.

CFCs were introduced in the 1930s and were used primarily as refrigerants and solvents. Large-scale use started in about 1950. Following the work of Molina, Rowland, and Crutzen in the 1970s, establishing the link between CFCs and stratospheric ozone depletion, and the discovery of an ozone hole over Antarctica in the mid-1980s, international agreements (protocols) on the use of CFCs were initiated. Use has been decreasing since the Montreal Protocol of 1987. This, and subsequent international agreements, instituted a phaseout of ozone-depleting chemicals. As shown in Figure 1-10, in some applications, CFCs have been replaced by hydrochlorofluorocarbons (HCFCs), which have shorter atmospheric lifetimes.

The partial destruction of the ozone layer in the stratosphere by the use and emission of CFCs is perhaps the best-documented example of the impact that human activities can have on the global environment, and the ability of nations to

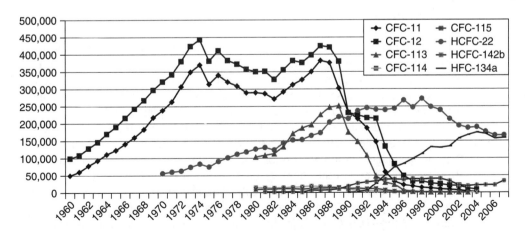

Figure 1-10 Recent trends in the production (in thousands of metric tons per year) of CFCs and HCFCs (AFEAS, 2009)

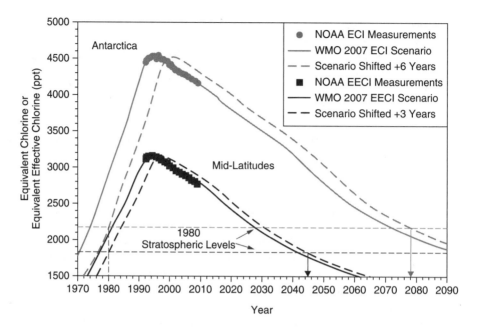

Figure 1-11   Recent trends in the tropospheric concentration of CFCs and solvents and scenarios for projected future concentrations (WMO is the World Meteorological Organization) (NOAA, 2010)

collectively respond to the challenge. Figures 1-10 and 1-11 show recent trends in the production of several CFCs and their remote tropospheric concentrations resulting from releases. The growth in accumulation of CFCs in the environment has been halted as a result of international agreements.

1.5.2   Global Warming

The atmosphere allows visible radiation from the sun to pass through without significant absorption of energy. Some solar radiation reaching the surface of the Earth is absorbed, heating the land and water. Infrared radiation is emitted from the Earth's surface, but certain gases in the atmosphere absorb this infrared radiation and redirect a portion back to the surface, thus warming the planet and making life, as we know it, possible. This process is often referred to as the *greenhouse effect*. The surface temperature of the Earth will rise until a radiative equilibrium is achieved between the rate of solar radiation absorption and the rate of infrared radiation emission (see Problem 7 at the end of the chapter). Human activities, such as fossil fuel combustion, deforestation, agriculture, and large-scale chemical production, have measurably altered the composition of gases in the atmosphere. The Intergovernmental Panel on Climate Change (IPCC, 2007a,b) has concluded that these alterations have led to a warming of the Earth-atmosphere system. Figure 1-12 summarizes the major links in the chain of environmental cause and effect for the emission of greenhouse gases.

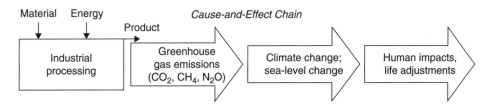

Figure 1-12   Greenhouse gas emissions and the major environmental cause-and-effect chain

Table 1-1 lists the most important greenhouse gases along with their anthropogenic (man-made) sources, emission rates, concentrations, residence times in the
atmosphere, relative radiative forcing efficiencies, and estimated contribution to global
warming. The primary greenhouse gases are water vapor, carbon dioxide, methane, nitrous oxide, CFCs, and tropospheric ozone. Water vapor is the most abundant greenhouse gas, but the concentration of water vapor in the atmosphere is not significantly
changed by anthropogenic emissions of water. Carbon dioxide contributes significantly to global warming because of its high emission rate, stability, and infrared
absorptivity. The major factors contributing to the global-warming potential of a
chemical are infrared absorptive capacity and residence time in the atmosphere. Gases
with very high absorptive capacities and long residence times can cause significant
global warming even though their concentrations are extremely low. A good example
of this phenomenon is the CFC class of chemicals, which are, on a pound-for-pound
basis, more than 1000 times more effective as greenhouse gases than carbon dioxide.

For the past five decades, measurements of the accumulation of carbon dioxide in the atmosphere have been taken at the Mauna Loa Observatory in Hawaii, a
location far removed from most human activity that might generate carbon dioxide.
Based on the current level of CO_2 of 390 ppm, levels of CO_2 are increasing at the
rate of 0.5% per year (from about 320 ppm in 1960). Atmospheric concentrations
of other greenhouse gases have also risen. Methane has increased from about
700 ppb in preindustrial times to 1790 ppb in 2009, and N_2O rose from 275 to 320
ppb over the same period. While it is clear that atmospheric concentrations of carbon dioxide and other global-warming gases are increasing, there is significant uncertainty regarding the magnitude of the effect on climate that these concentration
changes might induce (interested readers should consult the reports of the IPCC;
see the references at the end of the chapter). At the time of the writing of this book,
a variety of international protocols have been introduced with the goal of reducing
emissions of global warming gases.

1.5.3   Regional and Local Air Quality

Air pollution at regional and local scales arises from a number of sources, including
stationary, mobile, and area sources. Stationary sources include factories and other
manufacturing processes. Mobile sources are automobiles, trucks, other transportation

Table 1-1 Greenhouse Gases and Global Warming Contribution

Gas	Source (Natural and Anthropogenic)	Estimated Anthropogenic Emission Rate	Preindustrial Global Concentration	Approximate Current Concentration	Estimated Residence Time in Atmosphere	Radiative Forcing Efficiency over 100 Years ($CO_2 = 1$)	Estimated Contribution to Radiative Forcing (W/m^2)
Carbon dioxide (CO_2)	Fossil fuel combustion, deforestation	38 Gt/yr (10^{12} kg/yr)	280 ppm	390 ppm	50–200 yrs	1	1.5–1.8
Methane (CH_4)	Anaerobic decay (wetlands, landfills, rice paddies), ruminants, termites, natural gas, coal mining, biomass burning	0.3 Gt/yr	0.8 ppm	1.8 ppm	10 yrs	25	0.43–0.53
Nitrous oxide (N_2O)	Estuaries and tropical forests, agricultural practices, deforestation, land clearing, low-temperature fuel combustion	0.01 Gt/yr	0.385 ppm	0.32 ppm	140–190 yrs	298	0.14–0.18
Chlorofluoro-carbons	Refrigerants, air conditioners, foam blowing agents, aerosol cans, solvents	<0.0005 Gt/yr	0	0.0004–0.001 ppm	65–110 yrs	4750 (CFC-11)	0.31–0.37
Tropospheric ozone (O_3)	Photochemical reactions between VOCs and NOx from transportation and industrial sources	Not emitted directly	NA	0.022 ppm	Hours–days	2000	0.25–0.65

Source: Phipps, 1996; IPCC 2007a,b

vehicles, mobile construction equipment, and recreational vehicles such as snowmobiles and watercraft. Area sources are emissions associated with activities that are not considered mobile or stationary sources and that are associated with human activity. Examples of area sources include emissions from lawn and garden equipment and residential heating. Pollutants can be classified as primary, those emitted directly to the atmosphere, or secondary, being formed in the atmosphere after emission of precursor compounds. Photochemical smog (the term originated as a contraction of *smoke* and *fog*) is an example of secondary pollution that is formed from the emission of volatile organic compounds (VOCs) and nitrogen oxides (NOx), the primary pollutants. Air quality problems are closely associated with combustion processes occurring in the industrial and transportation sectors of the economy. In addition, hazardous air pollutants, including chlorinated organic compounds and heavy metals, are emitted in large enough quantities to be of concern. Figure 1-13 shows the main environmental cause-and-effect chain leading to the formation of smog.

Criteria Air Pollutants

Air pollutants are categorized in a variety of ways in U.S. regulations. One category is referred to as *criteria air pollutants*. The Clean Air Act Amendments of 1970 charged the United States Environmental Protection Agency (EPA) with identifying those air pollutants that most affect public health and welfare, and with setting maximum allowable ambient air concentrations (criteria) for these air pollutants. Six chemical species or groups of species (Table 1-2) have both primary and secondary standards that make up the National Ambient Air Quality Standards (NAAQS). The primary standards are intended to protect the public health with an adequate margin of safety. The secondary standards are meant to protect public welfare, such as damage to crops, vegetation, and ecosystems or reductions in visibility.

Since the establishment of the NAAQS, overall emissions of criteria pollutants in the United States have decreased even as population and economic activity have increased (a source of data is the *National Air Quality Emission Trends Report*, issued annually [U.S. EPA, 2010a]). Even with such improvements, in 2010 more than 100 million people in the United States lived in locations with air quality concentrations above the NAAQS (U.S. EPA, 2010a). The following sections briefly describe the criteria pollutants and their health effects.

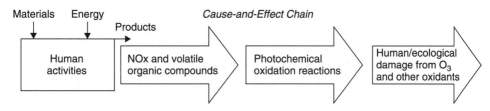

Figure 1-13   Environmental cause-and-effect chain for photochemical smog formation

Table 1-2 Criteria Pollutants and the National Ambient Air Quality Standards

Pollutant	Primary Standard (Human-Health-Related)		Secondary (Welfare-Related)	
	Type of Average	Concentration[a]	Type of Average	Concentration
CO	8-hour[b]	9 ppm (10 mg/m^3)	No Secondary Standard	
	1-hour[b]	35 ppm (40 mg/m^3)	No Secondary Standard	
Pb	Maximum quarterly average	1.5 μg/m^3	Same as Primary Standard	
	Rolling 3-month average	0.15 μg/m^3		
NO$_2$	Annual arithmetic mean	0.053 ppm (100 μg/m^3)	Same as Primary Standard	
	1-hour[c]	100 ppb		
O$_3$	1-hour[d]	0.12 ppm	Same as Primary Standard	
	8-hour[e]	0.075 ppm	Same as Primary Standard	
PM$_{10}$	24-hour[f]	150 μg/m^3	Same as Primary Standard	
PM$_{2.5}$	Annual arithmetic mean[g]	15 μg/m^3	Same as Primary Standard	
	24-hour[h]	35 μg/m^3	Same as Primary Standard	
SO$_2$	Annual arithmetic mean	0.03 ppm (80 μg/m^3)	3-hour[b]	0.5 ppm (1300 μg/m^3)
	24-hour[b]	0.14 ppm (365 μg/m^3)		
	1-hour[i]	75 ppb	None	

[a] Parenthetical value is an equivalent mass concentration.
[b] Not to be exceeded more than once per year.
[c] Three-year average of the annual 98th percentile of the daily maximum.
[d] Not to be exceeded more than once per year on average. The EPA has revoked this standard; however, some areas have continuing obligations under the standard.
[e] Three-year average of annual fourth-highest concentration. Not to be exceeded more than once per year on average over three years.
[f] Not to be exceeded more than once per year on average over three years.
[g] Three-year average of weighted means.
[h] The form is the 98th percentile.
[i] Three-year average of the annual 99th percentile of the daily maximum.
Source: Adapted from U.S. EPA (2011).

NOx, Hydrocarbons, and VOCs — Ground-Level Ozone

Ground-level ozone is one of the most pervasive and intractable air pollution problems in the United States. This "bad" ozone, created at or near ground level (tropospheric ozone), should again be differentiated from the "good" or stratospheric ozone that protects the surface of the planet from UV radiation (see Section 1.5.1). High ground-level ozone concentrations are exacerbated by certain physical and atmospheric factors. High-intensity solar radiation, low prevailing wind speed (low dilution), atmospheric inversions, and proximity to mountain ranges or coastlines (stagnant air masses) all contribute to photochemical smog formation. Human exposure to ozone can result in both acute (short-term) and chronic (long-term) health effects. The high reactivity of ozone makes it a strong lung irritant, even at low concentrations. Ground-level ozone also affects crops and vegetation when it enters the stomata of leaves and disrupts photosynthesis. Finally, ozone as an oxidant affects many materials such as painted surfaces and rubber.

Ground-level ozone is created in the atmosphere by precursor contaminants: VOCs and NOx (primarily NO and NO_2). The oxides of nitrogen along with sunlight cause ozone formation, but the role of VOCs is to accelerate and enhance the accumulation of O_3.

Oxides of nitrogen (NOx) are formed in high-temperature industrial and transportation combustion processes. The health effects of short-term exposure to NO_2 (less than 3 hours at high concentrations) are increases in respiratory illness in children and impaired respiratory function in individuals with preexisting respiratory problems. Between 1980 and 2008, NOx emissions have decreased by 40% in the United States (U.S. EPA, 2010a).

The VOC sources of most concern for smog formation are those that are easily oxidized in the atmosphere by the hydroxyl radical. Major sources of hydrocarbons are the chemical- and oil-refining industries and motor vehicles. It should be noted that there are natural (biogenic) sources of VOCs, such as isoprene and monoterpenes, that can contribute significantly to regional hydrocarbon emissions and ground-level ozone levels. Man-made emissions of VOCs decreased by 47% in the United States between 1980 and 2008 (U.S. EPA, 2010a).

Carbon Monoxide (CO)

CO is a colorless, odorless gas formed primarily as a by-product of incomplete combustion. The major health hazard posed by CO is its ability to bind with hemoglobin in the bloodstream and thereby reduce the oxygen-carrying ability of the blood. Transportation sources account for the bulk of total U.S. CO emissions. Ambient CO concentrations have decreased significantly in the past three decades, primarily because of vehicle exhaust emission controls and the introduction of oxygenated fuel additives that promote more complete combustion. Areas with high traffic congestion generally have high ambient CO concentrations. High localized and indoor CO levels can come from cigarettes (secondhand smoke), wood-burning fireplaces, and kerosene space heaters.

Lead

Lead in the atmosphere is primarily found in fine particles, up to 10 μ in diameter, which can remain suspended in the atmosphere for significant periods of time. Tetraethyl lead (($CH_3CH_2)_4$-Pb) was used as an octane booster and antiknock compound for many years before the introduction of automotive catalytic converters, driven by provisions of the Clean Air Act of 1970, forced the phaseout of all lead additives. The dramatic decline in lead concentrations and emissions due to the phaseout of lead additives in gasoline has been one of the most important environmental improvements of the past 30 years.

Particulate Matter (PM)

Particulate matter is the general term for micron-scale solid- or liquid-phase (aerosols) particles suspended in air. PM exists in a variety of sizes ranging from a few Angstroms to several hundred micrometers. Particles are either emitted directly from primary sources or are formed in the atmosphere by gas-phase reactions (secondary aerosols). Since particle size determines how deeply into the lung a particle is inhaled, there are two NAAQS for PM, $PM_{2.5}$ and PM_{10}. Particles smaller than 2.5 μm are called "fine" and are composed largely of inorganic salts (primarily ammonium sulfate and nitrate), sootlike carbon, organic species, and trace metals. Fine PM can deposit deep in the lung where removal is difficult. Particles larger than 10 μm are called "coarse" particles and are composed largely of suspended dust. Coarse PM tends to deposit in the upper respiratory tract, where removal is more easily accomplished. As with the other criteria pollutants, PM_{10} concentrations and emission rates have decreased because of pollution control efforts. Coarse particle inhalation is an irritant in upper respiratory difficulties. Fine particle inhalation can decrease lung functions, cause chronic bronchitis, and lead to premature mortality. Inhalation of specific toxic substances such as asbestos, coal mine dust, or textile fibers is now known to cause specific associated cancers (asbestosis, black lung cancer, and brown lung cancer, respectively).

An additional environmental effect of PM is limited visibility in many parts of the United States, including some National Parks. Also, nitrogen- and sulfur-containing particles are deposited, often with precipitation, increasing acidity levels in soil and water bodies and causing changes in soil nutrient balances and damage or death to aquatic organisms and ecosystems. PM deposition also causes soiling and corrosion of cultural monuments and buildings, especially those that are made of limestone.

SO_2, NOx, and Acid Deposition

Sulfur dioxide (SO_2) is the most commonly encountered of the sulfur oxide (SOx) gases and is formed upon combustion of sulfur-containing solid and liquid fuels (primarily coal and oil). SOx emissions are generated by transportation sources and electric utilities, metal smelting, and other industrial processes. Nitrogen oxides (NOx) are also produced in combustion reactions; however, the origin of most of the nitrogen is the combustion air rather than the fuel. After being emitted, SO_2 and NOx can be transported over long distances and are transformed in the atmosphere by gas-phase and aqueous-phase reactions to acid components (H_2SO_4 and HNO_3). The acid is deposited on the Earth's surface as either dry deposition of aerosols during periods of no precipitation or wet deposition of acid-containing rain or other precipitation. There are also natural emission sources for both sulfur- and nitrogen-containing compounds that contribute to acid deposition. Water in equilibrium with CO_2 in the atmosphere at a concentration of 390 ppm has a pH of 5 to 6. When natural sources of sulfur and nitrogen acid rain precursors are considered, the "natural" background pH of rain is expected to be about 5.0. As a result of these considerations, "acid rain" is defined as having a pH less than 5.0.

The main sources of SO_2 emissions are from non-transportation fuel combustion, industrial processes, and transportation. SO_2 concentrations and emissions decreased by approximately 60% between 1980 and 2008 (U.S. EPA, 2010a).

Emissions are expected to continue to decrease as a result of implementing the Acid Rain Program established by the EPA under the Clean Air Act. The goal of this program is to decrease acid deposition significantly by controlling SO_2 and emissions from utilities, smelters, and sulfuric acid manufacturing plants and reducing the average sulfur content of fuels for transportation and for industrial, commercial, and residential boilers.

There are a number of health and environmental effects of SO_2, NOx, and acid deposition. SO_2 is absorbed readily into the moist tissue lining the upper respiratory system, leading to irritation and swelling of this tissue and airway constriction. Long-term exposure to high concentrations can lead to aggravation of cardiovascular disease and lung disease. Acid deposition causes acidification of surface water, especially in regions of high SO_2 concentrations and low buffering and ion exchange capacity of soil and surface water. Acidification of water can cause harm to fish populations, by mobilization and exposure to heavy metals. Excessive exposure of plants to SO_2 decreases plant growth and yield and decreases the number and variety of plant species in a region.

Air Toxics

Hazardous air pollutants (HAPs), or air toxics, are airborne pollutants that are known to have adverse human health effects, such as cancer. Currently, more than 180 chemicals are identified on the Clean Air Act list of HAPs. Examples of air toxics are heavy metals like mercury and hexavalent chromium; and organic chemicals such as benzene, perchloroethylene (PERC), 1,3-butadiene, dioxins, and polycyclic aromatic hydrocarbons (PAHs).

The Clean Air Act defined a major source of HAPs as a stationary source that has the potential to emit 10 tons per year of any one HAP on the list or 25 tons per year of any combination of HAPs. The Clean Air Act prescribes a very high level of pollution control technology for HAPs called MACT (Maximum Achievable Control Technology). Small area sources, such as dry cleaners, emit lower HAP tonnages but taken together are a significant source of HAPs. Emission reductions can be achieved by changes in work practices, such as material substitution and other pollution prevention strategies.

HAPs affect human health via inhalation or ingestion routes. Some HAPs can accumulate in the tissue of fish, and the extent of the contaminant increases up the food chain to humans. Some of these persistent and bioaccumulative chemicals are known or suspected carcinogens.

1.5.4 Summary of Air Quality

Air quality issues provide an example of the complexity of environmental emissions and impacts. Spatial scales range from local to global. Temporal scales range from hours to decades. The chemical species involved are varied. Phenomena are nonlinear. Pollutants in the atmosphere may not stay in the atmosphere; they can be deposited onto land and water, linking air quality to water quality and ecosystem health. The same characteristics apply to many other environmental emissions and impacts. The remainder of this section will not describe these emissions and impacts at the same level of detail as the atmospheric emissions and impacts. This is not an indication of lesser importance. Instead, it is merely a reflection of the need to keep this introduction concise. Readers interested in learning more about other emissions and impacts are referred to the U.S. EPA's *Report on the Environment* (2008).

1.5.5 Water Quality

The availability of freshwater in sufficient quantity and purity is vitally important in meeting human domestic and industrial needs. Though 70% of the Earth's surface is covered with water, the vast majority exists in oceans and is too saline to meet the needs of domestic, agricultural, or other uses. Of the total 1.36 billion km^3 of water on Earth, 97% is ocean water, 2% is locked in glaciers, 0.31% is stored in deep groundwater reserves, and 0.32% is readily accessible freshwater (4.2 million km^3). Freshwater is continually replenished by the action of the hydrologic cycle. Surface water evaporates to form clouds; precipitation returns water to the Earth's surface, recharging the groundwater by infiltration through the soil; and rivers return water to the ocean to complete the cycle. In the United States, freshwater use is divided among agricultural irrigation, electricity generation, public supply, industry, and rural uses (Solley et al., 1993). Groundwater resources meet about 20% of U.S. water requirements, with the remainder coming from surface water sources.

Contamination of surface and groundwater originates from two categories of pollution sources. Point sources are entities that release relatively large quantities of wastewater at a specific location, such as industrial discharges and sewer outfalls. Non-point sources include all of the remaining discharges, such as agricultural and urban runoff, septic tank leachate, and mine drainage.

Besides the industrial and municipal sources we typically think of in regard to water pollution, other significant sources of surface and groundwater contamination are from agricultural and forestry activity. Contaminants originating from agricultural activities include pesticides; inorganic nutrients such as ammonium, nitrate, and phosphate; and leachate from animal waste. Forestry practices involve disruption of the soil surface from road building and the movement of heavy machinery on the forest floor. This activity increases erosion of topsoil, especially on steep forest slopes. The resulting additional suspended sediment in streams and rivers can lead to light blockage, reduced primary production in streams, destruction of spawning grounds, and habitat disruption of fisheries.

Transportation-related sources also contribute to water pollution, especially in coastal regions where shipping is most active. The 1989 *Exxon Valdez* oil spill in Prince William Sound in the state of Alaska and the 2010 *Deepwater Horizon* oil rig blowout in the Gulf of Mexico are well-known cases that coated shorelines with crude oil over a vast area. However, routine discharges of petroleum from oil tanker operations in 2002 were on the order of 8 million barrels/yr (NRC, 2002), which is an annual volume approximately 32 times higher than the *Exxon Valdez* spill. Transportation activities can also be a source of non-point pollution as runoff from roads carries oil, heavy metals, and salt into nearby streams.

1.5.6 Wastes in the United States

There is no single source of national industrial waste data in the United States. Instead, the national industrial waste generation, treatment, and release picture is a composite derived from multiple sources of data. A major source of industrial waste data is the EPA, which compiles various national inventories in response to legislative statutes. A sampling of the many laws requiring data collection include the Clean Air Act, Resource Conservation and Recovery Act (RCRA), Superfund Amendments and Reauthorization Act (SARA), and the Emergency Planning and Community Right-to-Know Act (EPCRA). In addition to these federal government sources of data, private industry is involved in data compilation activities, for example, the American Chemistry Council and the American Petroleum Institute. Following is a list of a number of national industrial waste databases (U.S. DOE, 1991). Because of the many inventories and the fact that more than one data source (inventory) might contain the same waste data, the assembly of the national waste picture is difficult. However, from these data sources it is possible to identify the major industrial sectors involved and the magnitude of their contributions.

Non-Hazardous Solid Waste

U.S. Environmental Protection Agency. 1988. *Report to Congress: Solid Waste Disposal in the United States, Volumes I and II*. EPA/530-SW-88-011 and EPA/530-SW-88-011B.

Criteria Air Pollutants

U.S. EPA, Office of Air Quality Planning and Standards. *Air Trends*. www.epa.gov/airtrends.

U.S. EPA, Office of Air Quality Planning and Standards. *National Air Pollutant Emission Estimates*. Research Triangle Park, NC.

Hazardous Waste (Air Releases, Wastewater, and Solids)

Biennial Report System (BRS); available through TRK NET, Washington, DC.

U.S. EPA, Office of Solid Waste. *National Biennial Report of Hazardous Waste Treatment, Storage, and Disposal Facilities Regulated under RCRA*. Washington, DC.

U.S. EPA, Office of Solid Waste. *Report to Congress on Special Wastes from Mineral Processing*. Washington, DC.

U.S. EPA, Office of Solid Waste. *Report to Congress: Management of Wastes from the Exploration, Development, and Production of Crude Oil, Natural Gas, and Geothermal Energy, Vol. 1, Oil and Gas*. Washington, DC.

Toxics Chemical Release Inventory (TRI); available through National Library of Medicine, Bethesda, MD, and RTK NET, Washington, DC. *Toxics Release Inventory: Public Data Release*. www.epa.gov/tri.

U.S. EPA, Office of Water Enforcement and Permits. *Permit Compliance System*. Washington, DC.

Economic Aspects of Pollution Abatement

U.S. Department of Commerce, Bureau of the Census. *Manufacturers' Pollution Abatement Capital Expenditures and Operating Costs*. Washington, DC.

U.S. Department of the Interior, Bureau of Mines. *Minerals Yearbook, Vol. 1: Metals and Minerals*. Washington, DC.

U.S. Department of Commerce, Bureau of the Census. Census Series: *Agriculture, Construction Industries, Manufacturers-Industry, Mineral Industries*. Washington, DC.

Non-hazardous industrial waste represents the largest contribution to national industrial waste generation. Roughly 12 billion tons of non-hazardous waste is generated and disposed of by U.S. industry (Allen and Rosselot, 1997; U.S. EPA, 1988a,b). That amount is over 200 pounds of industrial waste per person each day. This amount is about 60 times higher than the rate of waste generation by households in the United States (municipal solid waste). The largest industrial contributors to non-hazardous waste are manufacturing industries (7600 million tons/yr), oil and gas production (2095–3609 million tons/yr), and the mining industry (>1400 million tons/yr). Contributors of lower amounts are electricity generators (fly ash and flue-gas desulfurization waste), construction waste, hospital infectious waste, and waste tires.

Hazardous waste is defined under the provisions of the RCRA as residual materials having greater than a threshold value of ignitability, reactivity, toxicity, or corrosivity. Once this material is designated as hazardous, the costs of managing, treating, storing, and disposing of it increase dramatically. The rate of industrial hazardous waste generation in the United States is approximately 750 million tons/yr (Baker and Warren, 1992; Allen and Rosselot, 1997). This rate is 1/16 the rate of non-hazardous waste generation by industry. Furthermore, hazardous waste contains over 90% by weight of water, having only a relatively minor fraction of hazardous components. Therefore, the rate of generation of hazardous components in waste by industry is likely about 10 to 100 million tons/yr, though there is significant uncertainty in the exact amount because of differing definitions of hazardous waste. More than 90% of hazardous wastes are managed on-site at the facilities that generated them, and most of the hazardous waste is managed using wastewater treatment. Relatively little recycling and recovery of hazardous waste components occur in the current distribution of management methods.

Toxic releases from industrial operations are a subset of industrial wastes, but because of their potential to adversely affect human health and the health of the environment, they are reported separately. Approximately 650 of the thousands of chemicals used in commerce are reported through the U.S. EPA in the *Toxic Chemical Release Inventory* (TRI). Manufacturing operations and certain federal facilities are required to report to the TRI. Facilities must report releases of toxic chemicals to the air, water, and soil, as well as transfers to off-site recycling or treatment, storage, and disposal facilities. The release rate estimates include only the toxic components of any waste stream; thus water and other inerts are not included, in contrast with the industrial hazardous waste reporting system. The total releases and transfers reported to the TRI in 2008 (U.S. EPA, 2010b) were 2 million tons. Details of the releases are available from *TRI Trends*, published by the U.S. EPA (2010b).

1.6   SUMMARY

Engineers in the 21st century will need to design for energy efficiency, mass efficiency, and low environmental emissions. Figure 1-14 shows the types of transitions that must occur to achieve these goals (NRC, 2005). Both near- and long-term steps are needed reduce fossil resource consumption and approach zero waste generation from engineered processes and products.

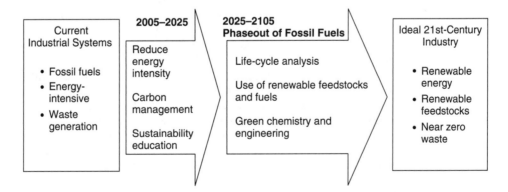

Figure 1-14   Grand challenges for a sustainable chemical industry (Adapted from NRC, 2005)

Some of these engineering challenges (e.g., energy efficiency) have always been a goal of engineering design. For other challenges (e.g., reducing emissions), the goals are newer, and these new challenges will require new types of tools. The engineers of the 21st century will need to adapt to these new demands and the use of these new tools.

[handwritten annotation: to convert next to electric 3412 Btu/kwh]

PROBLEMS

1. **The IPAT equation** Use the IPAT equation to estimate the percentage increase in the amount of energy that would be required, worldwide, in 2050, relative to 2006. To estimate the increase in population and affluence (the P and A in the IPAT equation), assume that population grows 1% per year and that global economic activity per person grows 2% per year. Assume that the energy consumption per dollar of GDP (the T in the IPAT equation) remains at 2006 levels. How much does this estimate change if population growth is 2% and economic growth is 4%?

2. **Affluence and energy use** Estimate the amount of energy that will be used annually, worldwide, if over the next 50 years world population grows to 10 billion and energy use per capita increases to the current per capita consumption rate in the United States. What percentage increase does this represent over current global energy use?

3. **Energy efficiency in automobiles** Assume that the conversion of energy into mechanical work (at the wheel) in an internal combustion engine is 20%. Calculate the gallons of gasoline required to deliver 30 horsepower at the wheel, for one hour.

4. **Water use by automobiles** Assuming that generating a kilowatt-hour of electricity requires an average of 13 gal of water (Example 1-3) and that an average electric vehicle requires 0.3 kWh/mi traveled (Kintner-Meyer et al., 2007), calculate the water use per mile traveled for an electric vehicle. If gasoline production requires approximately 10 gallons of water per gallon produced and an average gasoline-powered vehicle has a fuel efficiency of 25 mpg, calculate the water use per mile traveled of a gasoline-powered vehicle.

5. **Energy efficiency in lighting** Assume that a 25-watt fluorescent bulb provides the same illumination as a 100-watt incandescent bulb. Calculate the mass of coal that would be required, over the 8000-hour life of the fluorescent bulb, to generate the additional electricity required for an incandescent bulb. Assume transmission losses of 10%, 40% efficiency of electricity generation, and 10,000 BTU/lb for the heat of combustion of coal.

6. **Energy savings potential of compact fluorescent versus incandescent lightbulbs** Compact fluorescent lightbulbs provide similar lighting characteristics to incandescent bulbs yet use just one-fourth of the energy. Estimate the energy savings potential on a national scale of replacing all incandescent bulbs in home (residential) lighting applications with compact fluorescent bulbs. In 2008, total U.S. energy consumption was 99.3 quadrillion (10^{15}) BTU (quads), and electricity in all applications consumed 40.1 quads of *primary* energy. Assume that residential lighting is 3% of all electricity consumption in the United States and that all energy consumption for residential lighting is due to incandescent bulb use. How large (%) is the energy savings compared to annual U.S. energy consumption (2008 reference year)? Is this savings significant?

7. **Global energy balance: no atmosphere** (adapted from Wallace and Hobbs, 1977)
   a. The figure below is a schematic diagram of the Earth in radiative equilibrium with its surroundings, assuming no atmosphere. Radiative equilibrium requires that the rate of radiant (solar) energy absorbed by the surface must equal the rate of radiant energy emitted (infrared). Let S be the incident solar irradiance (1360 W/m²), E the infrared planetary irradiance (W/m²), R_E the radius of the Earth (meters), and A the planetary albedo (0.3). The albedo is the fraction of total incident solar radiation reflected back into space without being absorbed.

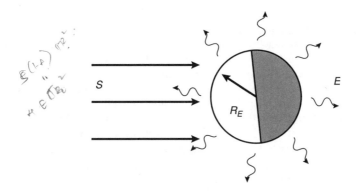

   b. Write the steady-state energy balance equation, assuming radiative equilibrium as stated above. Solve for the infrared irradiance, E, and show that its value is 238 W/m².
   c. Solve for the global average surface temperature (K), assuming that the surface emits infrared radiation as a black body. In this case, the Stefan-Boltzman Law for a black body is $E = \sigma T^4$, σ is the Stefan-Boltzman constant (5.67×10^{-8} W/(m²•°K⁴)), and T is absolute temperature (°K). Compare this temperature with the observed global average surface temperature of 280 K. Discuss possible reasons for the difference.

8. **Global energy balance: with a greenhouse gas atmosphere** (adapted from Wallace and Hobbs, 1977) Refer to the schematic diagram below for energy balance calculations on the atmosphere and surface of the Earth. Assume that the atmosphere can be regarded as a thin layer with an absorbtivity of 0.1 for solar radiation and 0.8 for infrared radiation. Assume that the Earth's surface radiates as a black body (absorbtivity = emissivity = 1.0).

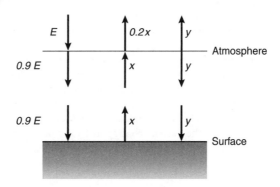

Let x equal the irradiance (W/m^2) of the Earth's surface and y the irradiance (both upward and downward) of the atmosphere. E is the irradiance entering the Earth-atmosphere system from space averaged over the globe ($E = 238$ W/m^2 from Problem 7). At the Earth's surface, a radiation balance requires that

$$0.9E + y = x$$
(irradiance in = irradiance out)

while for the atmosphere layer, the radiation balance is

$$E + x = 0.9E + 2y + 0.2x$$

   a. Solve these equations simultaneously for y and x.
   b. Use the Stefan-Boltzman Law (see Problem 7) to calculate the temperatures of both the surface and the atmosphere. Show that the surface temperature is higher than when no atmosphere is present (Problem 7).
   c. The emission into the atmosphere of infrared-absorbing chemicals is a concern for global warming. Determine by how much the absorbtivity of the atmosphere for infrared radiation must increase in order to cause a rise in the global average temperature by 1°C above the value calculated in Part b.

9. **Global carbon dioxide mass balance** Recent estimates of carbon dioxide emission rates to and removal rates from the atmosphere result in the following schematic diagram (EIA, 1998):

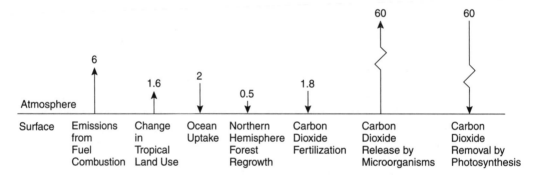

The numbers in the diagram have units of 10^9 metric tons of *carbon* per year, where a metric ton is equal to 1000 kg. To calculate the emission and removal rates for *carbon dioxide*, multiply each number by the ratio of molecular weights (44 g CO_2/12 g C).
   a. Write a steady-state mass balance for carbon dioxide in the atmosphere and calculate the rate of accumulation of CO_2 in the atmosphere in units of kilograms per year. Is the accumulation rate positive or negative?
   b. Change the emission rate due to fossil fuel combustion by +10% and recalculate the rate of accumulation of CO_2 in the atmosphere in units of kilograms per year. Compare this to the change in the rate of accumulation of CO_2 in the atmosphere due to a +1% change in carbon dioxide release by microorganisms.
   c. Calculate the rate of change in CO_2 concentration in units of parts per million per year, and compare this number with the observed rate of change stated in

Section 1.3.2. Recall the definition of parts per million (ppm), which for CO_2 is the mole fraction of CO_2 in the air. Assume that we are considering only the first 10 km in height of the atmosphere and that its gases are well mixed. Take for this calculation that the total moles of gas in the first 10 km of the atmosphere is approximately 1.5×10^{20}.

    d. Describe how the rate of accumulation of CO_2 in the atmosphere, calculated in Parts b and c, would change if processes such as carbon dioxide fertilization and forest growth increase as CO_2 concentrations increase. What processes releasing CO_2 might increase as atmospheric concentrations increase? (Hint: Assume that temperature will rise as CO_2 concentrations rise.)

10. Electric vehicles: effects on industrial production of fuels Replacing automobiles having internal combustion engines with vehicles having electric motors is seen by some as the best solution to urban smog and tropospheric ozone. Write a short report (one to two pages double-spaced) on the likely effects of this transition on industrial production of fuels. Assume for this analysis that the amount of energy required per mile traveled is roughly the same for each kind of vehicle. Consider the environmental impacts of using different kinds of fuel for the electricity generation to satisfy the demand from electric vehicles. Background reading for this problem is found in Graedel and Allenby (1998).

11. Essay on an environmental issue Read an article from a science or engineering journal, from a popular magazine, or from the Internet on some environmental issue that is of interest to you. Summarize the article in a short memorandum format, addressed to your instructor. In the body of the memorandum, limit the length to *one page* of single-spaced text, including graphics/tables (if needed). Structure the memorandum in this way:

    a. Introduction and motivation

    b. A description of the issue

    c. A description of what engineers are doing, have done, or are going to do to address the challenge

        Use of headings is appropriate, and be sure to reference information sources.

Potential topics:

- Stratospheric ozone depletion: the chemical industry connection
- Smog in industrialized urban areas
- Toxic chemicals in commerce and in the environment
- Industrial hazardous waste generation and management
- Environmental challenges for genetically engineered foods
- The cleanup of industrial sites (Superfund program)
- Pollution prevention issues, technologies, or initiatives
- Endocrine disruptors: What are they, why are they harmful, and what is the chemical industry doing about them?
- Environmental effects (advantages/disadvantages) of biodiesel or corn/cellulosic ethanol for transportation fuels
- Fuel cells and their environmental consequences
- Water resources: quality and quantity
- Petroleum: Are we running out? What are the alternatives?
- Renewable energy: What is it, and can it make a difference?

Potential sources of information:

Scientific and engineering research journals (check the library current journals section):

- *Environmental Science & Technology*
- *Environmental Progress and Sustainable Energy*
- *Industrial and Engineering Chemistry Research*
- *Chemical and Engineering News*
- *Science*
- *Scientific American*

Internet resources:

- American Chemistry Council (formerly the Chemical Manufacturers Association)
- U.S. Environmental Protection Agency (www.epa.gov)
- Your state's Department of Environmental Quality

12. **Sustainable development** An overview of the *Report of the World Commission on Environment and Development* is at www.un-documents.net/ocf-ov.htm. A number of global challenges to the environment, economic development, and living conditions were discussed. Summarize one or two of the key challenges in a memo format in one or two pages.

13. **International trade in waste** One of the consequences of the globalization of trade is a growing global trade in waste, often toxic and hazardous waste. Read a recent article on this subject and write a one- to two-page memo on key findings. Include a reference or references in your memo. One possible source of information is the International Network for Environmental Compliance and Enforcement (www.inece.org/seaport/SeaportWorkingPaper_24November.pdf).

REFERENCES

Adriaanse, A., S. Bringezu, A. Hammond, Y. Moriguchi, E. Rodenburg, D. Rogich, and H. Schηtz. 1997. *Resource Flows: The Material Basis of Industrial Economies.* Washington, DC: World Resources Institute.

AFEAS. 2009. *Alternative Fluorocarbons Environmental Acceptability Study.* www.afeas.org/. Accessed January 2010.

Allen, D. T. 2004. "An Industrial Ecology: Material Flows and Engineering Design." In *Sustainable Development in Practice: Case Studies for Engineers and Scientists,* edited by Dr. Adisa Azapagic, Dr. Slobodan Perdan, and Professor Roland Clift. West Sussex, UK: John Wiley & Sons, Ltd.

Allen, D. T., and K. S. Rosselot. 1997. *Pollution Prevention for Chemical Processes.* New York: John Wiley & Sons.

Argonne National Laboratory. 2011. *Fuel Properties.* www.afdc.energy.gov/afdc/fuels/properties.html. Accessed March 2011.

Baker, R. D., and J. L. Warren. 1992. "Generation of Hazardous Waste in the United States." *Hazardous Waste & Hazardous Materials* 9, no. 1 (Winter):19–35.

Ehrlich, P., and J. Holdren. 1971. "Impact of Population Growth." *Science* 171:1212–17.

EIA (Energy Information Agency, U.S. Department of Energy). 1998. *International Energy Outlook 1998.* DOE/EIA-0484(98). April.

———. 2009a. *International Energy Annual.* Available at www.eia.doe.gov/iea/.

———. 2009b. *Annual Energy Review, 2008.* www.eia.doe.gov/aer/.

EIOLCA, Economic Input-Output Life Cycle Assessment. Model available from Carnegie Mellon University. Available at www.eiolca.net. Accessed March 2011.

Graedel, T. E., and B. R. Allenby. 1995. *Industrial Ecology.* Englewood Cliffs, NJ: Prentice Hall, p. 232.

———. 1998. *Industrial Ecology and the Automobile.* Upper Saddle River, NJ: Prentice Hall.

IPCC (Intergovernmental Panel on Climate Change). 2007a. *Climate Change 2007, Synthesis Report.* Available at www.ipcc.ch.

———. 2007b. *Climate Change 2007, Summary for Policymakers. A Report of Working Group 1 of the Intergovernmental Panel on Climate Change.* Available at www.ipcc.ch.

Kintner-Meyer, M., K. Schneider, and R. Pratt. 2007. *Impact Assessment of Plug-In Hybrid Vehicles on Electric Utilities and Regional US Power Grids. Part I: Technical Analysis.* Pacific Northwest National Laboratory.

Lawrence Livermore National Laboratory. 2004. *Water Flow Charts, 2000.* UCRL-TR-201457-00. https://e-reports-ext.llnl.gov/pdf/308364.pdf. Accessed September 2011.

———. 2010. *National Energy Flows, 2008.* Work performed for the U.S. Department of Energy. https://publicaffairs.llnl.gov/news/energy/energy.html. Accessed January 2010.

Molina, M. J., and R. S. Rowland. 1974. "Stratospheric Sink for Chlorofluoromethanes: Chlorine Atom-Catalyzed Destruction of Ozone." *Nature* 249:810–12.

NOAA (National Oceanic and Atmospheric Administration), Earth System Research Laboratory. 2010. *Ozone Depleting Gas Index.* Boulder, CO. www.esrl.noaa.gov/gmd/odgi/. January.

NRC (National Research Council). 2002. *Oil in the Sea. III: Inputs, Fates and Effects.* Washington, DC: National Academy Press.

———. 2005. *Sustainability in the Chemical Industry: Grand Challenges and Research Needs—A Workshop Report.* Committee on Grand Challenges for Sustainability in the Chemical Industry, ISBN: 0-309-09571-9.

Phipps, E. 1996. *Overview of Environmental Problems.* Ann Arbor, MI: National Pollution Prevention Center for Higher Education, University of Michigan.

Solley, W. B., R. R. Pierce, and H. A. Perlman. 1993. *Estimated Use of Water in the United States 1990.* U.S. Geological Survey Circular 1081. Washington, DC: U.S. Government Printing Office.

United Nations. 2007. *World Population Prospects: The 2006 Revision.* Accessed at United Nations Population Information Network. www.un.org/popin/data.html.

U.S. DOE (U.S. Department of Energy). 1991. *Characterization of Major Waste Data Sources.* DOE/CE-40762T-H2.

U.S. EPA (United States Environmental Protection Agency). 1988a. *Report to Congress: Solid Waste Disposal in the United States, Volume 1.* EPA/530-SW-88-011.

———. 1988b. *Report to Congress: Solid Waste Disposal in the United States, Volume 1.* EPA/530-SW-88-011B.

————. 2008. *Report on the Environment*. EPA/600/R-07/045F. May. Available at www.epa.gov/roe/.

————. 2010a. *Air Trends*. www.epa.gov/airtrends.

————. 2010b. *Toxics Release Inventory*. www.epa.gov/tri.

————. 2011. *National Ambient Air Quality Standards*. www.epa.gov/air/criteria.html.

USGS (United States Geological Survey). 1998. *Materials Flow and Sustainability, Fact Sheet*. FS-068-98. June. Available at http://pubs.usgs.gov/fs/fs-0068-98/fs-0068-98.pdf.

————. 2000. *Materials and Energy Flows in the Earth Science Century*. USGS Circular 1194.

————. 2002. *Materials in the Economy: Material Flows, Scarcity, and the Environment*. USGS Circular 1221. February.

Wallace, J. M., and P. V. Hobbs. 1977. *Atmospheric Science: An Introductory Survey*. New York: Academic Press.

World Commission on Environment and Development. 1987. *Our Common Future*. Oxford: Oxford University Press.

CHAPTER 2

Risk and Life-Cycle Frameworks for Sustainability[1]

2.1 INTRODUCTION

Sustainability issues are complex, and understanding their interactions with equally complex engineered systems is best done through well-structured analysis frameworks. Risk-based frameworks have traditionally been used in the characterization and prioritization of environmental issues, and the first section of this chapter describes those frameworks. Increasingly, however, life-cycle frameworks are gaining prominence as a means of characterizing and understanding sustainability. The second major section of this chapter introduces life-cycle frameworks and their uses.

2.2 RISK

Understanding risk frameworks for environmental and sustainability issues requires an understanding of the language of risk, the tools used to quantify risk, and the use of risk in regulation. This section addresses each of these prerequisites for understanding the use of risk frameworks.

2.2.1 Definitions

Risk is defined as the potential for an individual to suffer an adverse effect from an event. The concept of risk is used in many disciplines, such as finance, engineering,

---

1. The portion of this chapter describing risk frameworks and the portion describing life-cycle frameworks were adapted from Chapters 2 and 13 in the text *Green Engineering* (Allen and Shonnard, 2002). The authors thank their collaborators on these chapters in the *Green Engineering* text (Fred Arnold, John Blouin, Gail Froiman, and Kirsten Rosselot) for their contributions to the ideas presented in this chapter.

and health, and the precise definition of risk can vary, depending on the application. Environmental risks, which are the focus of this chapter, can be grouped into three general categories:

- **Risks involving voluntary exposure:** Activities done for a living or for enjoyment (firefighting, skydiving, mountain climbing, bungee cord jumping, etc.). The risk (danger) is usually obvious and the activity is usually done by free will.
- **Risks associated with natural disasters:** Floods, hurricanes, earthquakes, meteorite hits, and other disasters beyond human control. Exposure to the effects of certain natural disasters can be exacerbated by actions such as living on a known earthquake fault or the slope of a volcano.
- **Risks involving involuntary exposure:** An individual or entity releases a compound into the environment (pesticides, known carcinogens), potentially harming workers or members of the public, who cannot directly control the exposure.

The magnitudes of involuntary environmental risks are often quite different from the magnitudes of the risks associated with voluntary activities or natural disasters. Table 2-1 lists one assessor's evaluation of various risks.

In this chapter, the focus is on the concept of involuntary environmental risks associated with material, radiation, or other releases into the environment. Environmental risk is a function of hazard and exposure:

$$\text{Environmental risk} = f\,(\text{hazard, exposure})$$

The hazard depends on the toxicity of the material, radiation, or other release. Exposure depends on the concentration of the material that a person experiences. Engineers can reduce environmental risk by using less toxic material (minimizing hazard) or by engineering systems that do not allow materials to escape into the environment (minimizing exposure).

Environmental risk assessment is used to quantitatively determine the probability of the adverse effects of environmental releases. A common application of risk assessment methods is to evaluate human health and ecological impacts of chemical releases to the environment. Information collected from environmental monitoring or modeling is incorporated into models of human or worker activity, and estimates of the likelihood of adverse effects are formulated, as illustrated in Example 2-1. Risk assessment is an important tool for making decisions with environmental consequences. Almost always, when the results from environmental risk assessment are used, they are incorporated into the decision-making process along with the economic, societal, technological, and political consequences of a proposed action.

Example 2-1   Carcinogenic risk assessment near a petroleum refinery

A petroleum refinery is performing a quantitative risk assessment on the atmospheric releases of volatile organic compounds from the facility, some of which are toxic. Assess the risk of benzene released to the air from the facility based on its impact on

Table 2-1   Loss of Life Expectancy from Various Societal Activities and Phenomena

Risk Factor	Loss of Life Expectancy (Days)
Cancer Risks Associated with Environmental Pollutants	
Indoor radon	30
Worker chemical exposure	30
Pesticide residues in food	12
Indoor air pollution	10
Consumer product use	10
Stratospheric ozone depletion	22
Inactive hazardous waste sites	2.5
Carcinogens in air pollution	4
Drinking water contaminants	1.3
Noncancer Risks Associated with Environmental Pollutants	
Lead	20
Carbon monoxide	20
Sulfur dioxide	20
Radon	0.2
Air pollutants (e.g., carbon tetrachloride, chlorine, etc.)	0.2
Drinking water contaminants (e.g., lead, pathogens, nitrates, chlorine disinfectants, etc.)	0.2
Industrial discharge into surface water	Few minutes
Sewage treatment plant sludge	Few minutes
Mining wastes	Few minutes
Lifestyle/Demographic Status Risks	
Being an unmarried male	3500
Smoking cigarettes and being male	2250
Being an unmarried female	1600
Being 30% overweight	1300
Being 20% overweight	900
Having less than an eighth-grade education	850
Smoking cigarettes and being female	800
Being poor	700
Smoking cigars	330
Having a dangerous job	300
Driving a motor vehicle	207
Drinking alcohol	130
Having accidents in the home	95
Suicide	95
Being murdered	90
Misusing legal drugs	90

Source: Fan and Chang (1996), p. 247

human health (carcinogenic impact, inhalation only) in a hypothetical residential area downwind of the facility, assuming that the maximum average annual concentration of benzene in the *outside* air (C_A) within the residential area is 82 μg/m³. The dose-response carcinogenic slope factor (SF) for benzene inhalation is 2.9×10^{-2} (mg benzene/(kg body weight • day))$^{-1}$. The slope factor relates the quantity of benzene absorbed through the lungs to the fraction of the population that will contract cancer (increased probability of cancer = inhaled benzene, in mass per body weight per day *SF).

Use the following exposure properties:

BW:	average adult body weight = 70 kg
CR:	air breathing rate = 19.92 m³/day
RR:	retention rate, inhaled air = 1.0
ABS:	absorption rate, inhaled air = 1.0
EF:	exposure frequency = 365 exposure days/yr
ED:	exposure duration = 70 yr
AT:	averaging time = 25,550 days

where RR is the efficiency of the lungs to retain benzene and ABS is the efficiency of the lung tissue to absorb the retained chemical. These values were to set to maximum values (1.0) for this problem and may actually be much lower.

a. Calculate the inhalation dose of benzene to a typical resident using the following equation:

$$\text{Inhalation Dose (mg benzene/(kg body weight • day))} = \frac{C_A \times CR \times EF \times ED \times RR \times ABS}{BW \times AT}$$

b. Calculate the inhalation carcinogenic risk for this scenario using the following equation:

$$\text{Inhalation Carcinogenic Risk (dimensionless)} = \text{Inhalation Dose} \times SF$$

c. Is the risk greater than the recommended range of $<10^{-4}$ to 10^{-6} (1 in 10,000 to 1 in 1,000,000 additional people contracting cancer due to inhaled benzene) for carcinogenic risk?

d. Discuss possible reasons why this methodology might overpredict the actual risk.

Solution:

a. Inhalation Dose, I_{INH}:

$$\text{Inhalation Dose (mg benzene/(kg body weight • day))} = \frac{C_A \times CR \times EF \times ED \times RR \times ABS}{BW \times AT}$$

$$= (82 \text{ μg/m}^3) (10^{-3} \text{ mg/μg}) (19.92 \text{ m}^3\text{/day}) (365 \text{ day/yr}) (70 \text{ yr}) (1) (1) / [70 \text{ kg} * 25{,}550 \text{ day}]$$

$$= 0.023 \text{ mg benzene/[kg body weight-day]}$$

Note that the number of days in the numerator and denominator is the same, since it is assumed that the exposed individual remains in the same location, the point of maximum concentration, for 70 years.

b. Inhalation Carcinogenic Risk

$$\text{Inhalation Carcinogenic Risk} = I_{INH} \times SF$$

$$= \left(0.02331 \frac{\text{mg Benzene}}{\text{kg BW ¥day}}\right)\left(2.9 \times 10^{-2}\left[\frac{\text{mg Benzene}}{\text{kg BW ¥day}}\right]^{-1}\right)$$

$$= 6.76 \times 10^{-4}$$

c. Risk of cancer due to inhalation is greater than the recommended range of $<10^{-4}$ to 10^{-6}.

d. This calculation might overpredict the risk because RR and ABS were assumed to be 1 when their actual value would be less than 1 but greater than 0 (the bloodstream will not absorb all of the benzene inhaled into the lungs). Also, it was assumed that exposure to outside air, at the point of maximum concentration, would occur for 24 hours per day, 365 days per year, for 70 years. In practice, people move about over a 70-year lifetime.

2.2.2 Risk Assessment

In 1983, the National Research Council (NRC, 1983) developed an environmental risk assessment framework that, updated (NRC, 2009), is still in place today. The framework consists of four major components: hazard assessment, dose response, exposure assessment, and risk characterization.

1. **Hazard assessment:** What are the adverse health effects of the chemical(s) in question? Under what conditions? Toxicologists usually perform this analysis. Sources of hazard information are provided in the appendix to this chapter.

2. **Dose response:** How much of the chemical causes a particular adverse effect? Are some individuals more sensitive to a particular dose than others? There may be multiple adverse health effects, or responses, for the same chemical, and each adverse effect has a unique relationship between dose and response. For our purposes, dose is defined as the quantity of a chemical that crosses a boundary to get into a human body or organ system. The term applies regardless of whether the substance is inhaled, ingested, or absorbed through the skin. Dose response, then, is a mathematical relationship between the magnitude of a dose and the magnitude of a certain response in the exposed population.

3. **Exposure assessment:** Who is exposed to this chemical? How much of the chemical reaches the boundary of a person, and how much enters the person's body? Exposure may be measured, estimated from models, or even calculated from measurements of biomarkers taken from exposed individuals.

4. **Risk characterization:** How great is the potential for adverse impact from this chemical? What are the uncertainties in the analyses? How conclusive are the results of these analyses?

The risk assessment process can be iterative. That is, if a cursory or screening risk assessment identifies concerns, a more rigorous process may be called for. This process may in turn illustrate that there are important, specific data gaps that need to be filled to render the risk assessment process conclusive enough for risk management. The data gaps may be filled with recommendations for special studies of varying cost and time requirements, such as

- Proceeding with testing for health effects
- Evaluating the effectiveness of engineering controls, to limit exposure to chemicals, and of personnel protective equipment
- Defining the degradation kinetics and decomposition products of a waste stream and the impact of the chemical waste and its degradation products on local flora and fauna

If it is reasonably clear from the risk assessment that a risk exists, the next step is risk management.

Risk management is the process of identifying, evaluating, selecting, and implementing actions to reduce risk to human health and to ecosystems. The goal of risk management is scientifically sound, cost-effective, integrated actions that reduce or prevent risks while taking into account social, cultural, ethical, political, and legal considerations (Presidential Commission, 1997).

Risk managers must answer many questions, some of which are:

- What level of exposure to a chemical risk agent is an unacceptable risk?
- How great are the uncertainties and are there any mitigating circumstances?
- Are there any trade-offs between risk reduction, benefits, and additional cost?
- What are the chances of risk shifting, that is, shifting risk to other populations?
- Are some of the risks worse than others?

The answers to these questions can depend on culture and values, as illustrated by the use of risk assessment in environmental law.

2.2.3 Risk-Based Environmental Law

Many of our environmental laws are based on risk frameworks. Table 2-2 lists selected U.S. safety, health, and environmental statutes that require or suggest human health risk assessment before regulations are promulgated. These regulatory risk assessments are made complex because not all environmental statutes (laws) are developed using the same types of standards. For example, the provisions of the Clean Air Act pertaining to National Ambient Air Quality Standards (NAAQS) call for values that "protect the public health allowing an adequate margin of safety." That is, these standards mandate protection of public health based only on risk, without regard to technology or cost factors. In contrast, the Clean Water Act requires industries to install specific treatment technologies. These have descriptions like "best practicable control technology" and "best available technology economically achievable." Pesticides are licensed if they do not cause "any unreasonable risks to man or the environment taking into account the economic, social, and environmental costs and benefits of the use of any pesticide." In this case, economic and other factors may or may not be combined with risk issues as regulations are developed. So, the details of the risk frameworks used in environmental regulations vary, but the basic principles of the framework remain the same.

Section 812 of the Clean Air Act Amendments of 1990 provides a case study of a statute that requires a particularly detailed risk assessment. The goal of this section of the Clean Air Act is to assess the costs and the benefits of the Clean Air Act. It requires the Environmental Protection Agency to estimate the hazards associated with air pollutants covered under the act, use dose-response curves to estimate health effects (and ecological effects to the extent possible), use exposure estimates to determine health impacts in the U.S. population, characterize uncertainties, and perform a form of risk management (cost-benefit analysis) (for more details, see U.S. EPA, 2011a). These steps are outlined here:

- **Hazard identification:** Mortality and disease associated with exposure to air pollutants are the primary hazards considered in Section 812 analyses; some risks to ecosystems, such as the nitrification of bays, have been considered on a case-by-case basis.

- **Dose response:** Quantitative relationships are developed for increased mortality and disease associated with exposure to air pollutants; the primary analyses are for exposure to particulate matter and ground-level ozone.

- **Exposure assessment:** The exposure of the entire U.S. population to these pollutants is estimated, based on geographical variability of pollutant concentrations (for example, ozone concentrations are higher in Los Angeles than in Houghton, Michigan).

- **Risk characterization:** Mortality and the prevalence of disease caused by ozone and particulate matter are estimated; uncertainties are estimated and reported; and the estimated reductions in mortality and disease due to the regulations in the Clean Air Act are estimated.

- **Risk management:** To help inform decisions about whether Clean Air Act regulations are too strict, not strict enough, or at about the right level, the costs and benefits of the regulations are estimated. Costs are expressed in dollars and are estimated based on the costs of equipment (e.g., exhaust controls on vehicles) and the costs of operations (e.g., energy needed to run some types of pollution control devices such as electrostatic precipitators to capture particles). The benefits of the regulations are avoided early deaths (death cannot be avoided, but it can be hastened) and avoided disease. The risk manager could make decisions based on the estimates of monetary costs of the controls, weighed against avoided early deaths and avoided disease, but this would involve an implicit evaluation of the monetary value of disease and early death. An interesting feature of the Section 812 analyses performed by EPA is that the benefits of the regulation are explicitly monetized. The benefit of disease reduction is monetized by calculating direct health care costs associated with treating the diseases. Monetizing avoided early deaths is more problematic. One approach to monetizing the costs of early deaths has been to estimate lost wages due to decreased life expectancy. This means that the life of a well-educated college student is worth more than the life of that student's grandparent, because if the student were to die prematurely because of air pollution, the lifetime loss in expected wages would be greater than if the grandparent were to die early from air pollution exposure. But is one life worth more than another? Currently, in estimating the monetary benefit of avoided early mortality, the EPA values all avoided early deaths equally—at about $7 million to $9 million. These assumptions lead to estimated benefits of the Clean Air Act (about $2 trillion) that far outweigh the estimated costs (about $85 billion) (U.S. EPA, 2011b).

2.3   LIFE-CYCLE FRAMEWORKS

Understanding life-cycle frameworks for environmental and sustainability issues requires an understanding of the language of life-cycle assessment, the tools used to quantify life-cycle impacts, and the use of life-cycle assessment in regulation. This section addresses each of these prerequisites for understanding the use of life-cycle frameworks and provides case studies of the use of life-cycle assessment tools.

2.3.1   Defining Life Cycles

While risk assessment and risk management have proven to be useful frameworks for managing environmental issues, one shortcoming of these approaches is that they tend to examine individual events (e.g., a single pollutant being emitted from a single source or group of sources) rather than systems. Decisions made to address environmental issues or sustainability often have cascading impacts through complex engineered systems. Consider, as an example, the use of biofuels to reduce the net emissions of greenhouse gases.

Burning a biofuel (such as ethanol derived from sugarcane or corn) releases carbon dioxide, but that carbon was originally withdrawn by the biomass from the atmosphere during the process of photosynthesis. When the atmospheric withdrawal due to photosynthesis is combined with the combustion emissions, the net emissions of carbon dioxide to the atmosphere are lowered for biofuels. In contrast, burning petroleum-based fuels releases carbon to the atmosphere that was originally (at least over the timescale of centuries) sequestered in geological formations. As a consequence, the use of biofuels is often seen as a strategy for reducing net greenhouse gas emissions.

Table 2-2   U.S. Safety, Health, and Environmental Statutes That Imply Risk Assessment

Agency	Act	U.S. Code
United States Environmental Protection Agency	Atomic Energy Act (also NRC)	42 U.S.C. 2011
	Comprehensive Environmental Response, Compensation, and Liability Act (CERCLA, or Superfund)	42 U.S.C. 9601
	Clean Air Act	42 U.S.C. 7401
	Clean Water Act	33 U.S.C. 1251
	Emergency Planning and Community Right-to-Know Act	42 U.S.C. 11001
	Federal Food, Drug, and Cosmetics Act (also HHS)	21 U.S.C. 301
	Federal Insecticide, Fungicide, and Rodenticide Act	7 U.S.C. 136
	Lead Contamination Control Act of 1988	42 U.S.C. 300j-21
	Marine Protection, Research, and Sanctuaries Act (also DA)	16 U.S.C. 1431
	Nuclear Waste Policy Act	42 U.S.C. 10101
	Resource Conservation and Recovery Act	42 U.S.C. 6901
	Safe Drinking Water Act	42 U.S.C. 300f
	Toxic Substances Control Act	7 U.S.C. 136
	Food Quality Protection Act of 1996	7 U.S.C. 6
Consumer Product Safety Commission	Consumer Product Safety Act	15 U.S.C. 2051
	Federal Hazardous Substance Act	15 U.S.C. 1261
	Lead-Based Paint Poisoning Act (also HHS and HUD)	42 U.S.C. 4801
	Lead Contamination Control Act of 1988	42 U.S.C. 300j-21
	Poison Prevention Packaging Act	15 U.S.C. 1471
Department of Agriculture	Egg Products Inspection Act	21 U.S.C. 1031
	Federal Meat Inspection Act	21 U.S.C. 601
	Poultry Products Inspection Act	21 U.S.C. 451
Department of Labor	Federal Mine Safety and Health Act	30 U.S.C. 801
	Occupational Safety and Health Act	29 U.S.C. 651
Department of Transportation	Hazardous Liquid Pipeline Safety Act	49 U.S.C. 1671
	Hazardous Materials Transportation Act	49 U.S.C. 1801
	Motor Carrier Safety Act	49 .U.S.C. 2501
	National Traffic and Motor Vehicle Safety Act	15 U.S.C. 1381
	National Gas Pipeline Safety Act	49 U.S.C. 2001

Source: Federal Focus, 1991; Roberts and Abernathy, 1996; Fort, 1996

While use of biofuels may reduce net carbon dioxide emissions to the atmosphere if just photosynthesis and combustion are considered, biofuels are the product of a complex engineered system. Growing biomass for biofuels requires fertilizer. Fertilizers require energy to manufacture and may cause the release of N_2O (a potent greenhouse gas) when applied to soils. Growing biomass for biofuels requires land. As land is converted from nonagricultural uses into agricultural production, carbon in the soil and in the original vegetation covering the land may be lost. Growing biomass for biofuels sometimes requires irrigation. Pumping water requires energy. So, the decision to grow biomass for biofuels will change a complex agricultural system in complicated ways. Frameworks are needed to understand these systems. Life-cycle frameworks are one possibility.

Products, services, and processes all have a life cycle. For products, the life cycle begins when raw materials are extracted or harvested. Raw materials then go through a number of manufacturing steps until the product is delivered to a customer. The product is used, then disposed of or recycled. These product life-cycle stages are illustrated in Figure 2-1, along the horizontal axis. As shown in the figure, energy is consumed, and wastes and emissions are generated in all of these life-cycle stages.

Processes also have a life cycle. The life cycle begins with planning, research, and development. The products and processes are then designed and constructed. A process has an active lifetime, then is decommissioned, and, if necessary, remediation and restoration may occur. Figure 2-1, along its vertical axis, illustrates the main elements of this process life cycle. Again, energy consumption, wastes, and emissions are associated with each step in the life cycle.

Traditionally, product designers have been concerned primarily with product life cycles up to and including the manufacturing step. Process designers have been primarily concerned with process life cycles up to and including the manufacturing step. That focus is changing in the design process. Increasingly, product designers must consider how their products will be recycled. They must consider how their customers will use their products and what environmental hazards might arise. Process designers must avoid contamination of the sites where their processes are located. Simply stated, engineers must become stewards for their products and processes throughout their life cycles.

2.3.2 Life-Cycle Assessment

There is some variability in life-cycle assessment terminology, but the most widely accepted terminology was originally codified by international groups convened by the Society for Environmental Toxicology and Chemistry (SETAC) (see, for example, Consoli et al., 1993) and has since been codified in the 14040 series of life-cycle assessment standards, issued by the International Standards Organization (ISO, 2006). To begin, a *life-cycle assessment* (LCA) is the most complete and detailed form of a life-cycle study. A life-cycle assessment consists of four major steps.

Step 1. The first step in an LCA is to determine the scope and boundaries of the assessment. In this step, the reasons for conducting the LCA are identified; the

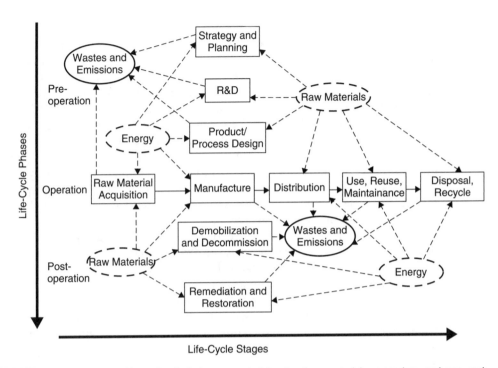

Figure 2-1 Product life cycles include raw material extraction, material processing, and use and disposal steps and are illustrated along the horizontal axis. Process life cycles include planning, research, design, operation, and decommissioning steps and are shown along the vertical axis. In both product and process life cycles, energy and materials are used at each stage, and emissions and wastes are created.

product, process, or service to be studied is defined; a functional unit for that product is chosen; and choices regarding system boundaries, including temporal and spatial boundaries, are made. But what is a functional unit, and what do we mean by system boundaries? Let's look first at system boundaries.

The *system boundaries* are simply the limits placed on data collection for the study. The importance of system boundaries can be illustrated by a simple example. Consider the problem of choosing between incandescent lightbulbs and fluorescent lamps for lighting a room. During the 1990s the U.S. EPA began its Green Lights Program, which promoted replacing incandescent bulbs with fluorescent lamps. The motivation was the energy savings provided by fluorescent bulbs. Like any other product, however, a fluorescent bulb is not completely environmentally benign, and a concern arose during the Green Lights Program about the use of mercury in fluorescent bulbs. Fluorescent bulbs provide light by causing mercury, in glass tubes, to fluoresce. When the bulbs reach the end of their useful life, the mercury in the tubes might be released to the environment. This environmental concern (mercury release during product disposal) is far less significant for incandescent bulbs. Or is it?

What if we changed our system boundary? Instead of just looking at product disposal, as shown in the first part of Figure 2-2, what if the entire product life cycle were considered, as shown in the second half of Figure 2-2? In a comparison of the incandescent and fluorescent lighting systems, if the system boundary is selected to include electric power generation as well as disposal, the analysis changes. Mercury is a trace contaminant in coal, and when coal is burned to generate electricity, some mercury is released to the atmosphere. Since an incandescent bulb requires more energy to operate, the use of an incandescent bulb results in the release of more mercury to the atmosphere than the use of a fluorescent bulb. Over the lifetime of the bulbs, more mercury can be released to the environment from energy use than from disposal of fluorescent bulbs. Thus, the simple issue of determining which bulb, over its life cycle, results in the release of more mercury depends strongly on how the boundaries of the system are chosen.

As this example illustrates, the choice of system boundaries can influence the outcome of a life-cycle assessment. A narrowly defined system requires less data collection and analysis but may ignore critical features of a system. On the other hand, in a practical sense it is impossible to quantify all impacts for a process or product system. Returning to our example, should we assess the impacts of mining the metals and making the glass used in the bulbs we are analyzing? In general, we would not need to consider these issues if the impacts are negligible, compared to

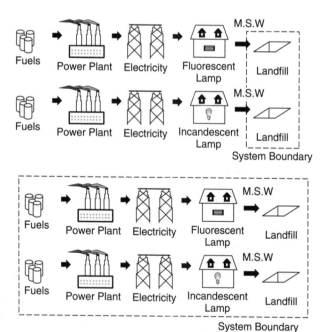

Figure 2-2 The importance of system boundaries in life-cycle assessment is illustrated by the case of lighting systems (see the text).

the impacts associated with operations over the life of the equipment. On the other hand, for specific issues, such as mercury release, some of these ancillary processes could be important contributors. What is included in the system and what is left out are often based on engineering judgment and a desire to capture any parts of the system that may account for 1% or more of the energy use, raw materials use, wastes, or emissions (other approaches to defining system boundaries are possible; see Allen et al., 2009).

Another critical part of defining the scope of a life-cycle assessment is to specify the *functional unit*. The choice of functional unit is especially important when life-cycle assessments are conducted to compare products. This is because functional units are necessary for determining equivalence between the choices. For example, if paper and plastic grocery sacks are to be compared in an LCA, it would not be appropriate to compare one paper sack to one plastic sack. Instead, the products should be compared based on the volume of groceries they can carry. Because fewer groceries are generally placed in plastic sacks than in paper sacks, some LCAs have assumed a functional equivalence of two plastic grocery sacks to one paper sack. Differing product lifetimes must also be evaluated carefully when using life-cycle studies to compare products. For example, a cloth grocery sack may be able to hold only as many groceries as a plastic sack but will have a much longer use during its lifetime that must be accounted for in performing the LCA. As shown in Example 2-2, the choice of functional unit is not always straightforward and can have a profound impact on the results of a study.

Example 2-2   Functional units

Propose functional units for comparing

    a. Paper and plastic grocery sacks
    b. Paper and cloth grocery sacks
    c. Transportation fuels
    d. Compact fluorescent, LED, and incandescent lightbulbs

Solution:

    a. The function of a grocery sack is to transport groceries from one location to another, so an appropriate functional unit would be the number of grocery sacks required to transport a well-defined basket of groceries. Most studies of grocery sacks have found that it takes roughly twice as many (2 ± 1) plastic sacks as paper sacks to perform the function of transporting groceries (Allen et al., 1992).
    b. Again, the function of a grocery sack is to transport groceries from one location to another, so an appropriate functional unit would factor in the number of grocery sacks required to transport a well-defined basket of groceries. Because of the different product lifetimes of the grocery sacks, however, an additional factor that should be considered in an appropriate functional unit is product lifetime. So, in this case, an appropriate functional unit could be the number of sacks needed over a large number (e.g., 1000) of visits to the grocery store. The number of times each product is reused would need to be considered, as well as any processing steps between uses (e.g., washing the bags).

c. The function of a transportation fuel is to provide combustion energy to an engine. The most common unit is megajoules (MJ) of lower heating value (LHV) provided on combustion. LHV is used instead of higher heating value because water in the exhaust of engines is in the gas phase.

d. For lighting devices, the most common functional unit is a quantity of illumination provided over a given amount of time. So, for example, if a 25-watt compact fluorescent (CFL) bulb lasts for 8000 hours of use and provides 1700 lumens of illumination, and a 100-watt incandescent bulb provides 1700 lumens of illumination and lasts for 1000 hours of use, then to provide 1700 lumens for 8000 hours requires 1 CFL bulb and 8 incandescent bulbs. The incandescent bulbs would use 800 kWh of electricity over this time period (100 watts * 8000 hours), while the CFL bulb would use 200 kWh (25 watts * 8000 hours). Making the situation more complex is the fact that bulbs produce both heat and light. A bulb operated indoors will both heat a room and provide light. The heat may be desirable on a cold winter day in Houghton, Michigan, but undesirable on a hot summer day in Austin, Texas. So, what is the function of the bulb? Providing light? Providing heat? Both?

Step 2. The second step in a life-cycle assessment is to inventory the outputs that occur, such as products, by-products, wastes, and emissions, and the inputs, such as raw materials and energy, that are used during the life cycle. This step, shown conceptually in Figure 2-3, is called a *life-cycle inventory* and is often the most time-consuming and data-intensive portion of a life-cycle assessment.

An example of the level of detail that can be associated with a life-cycle inventory is shown in Tables 2-3 and 2-4. The data, drawn from a publicly available database of life-cycle inventory information (NREL, 2011, described later in this chapter), shows the inputs required to produce a kilogram of corn and a kilogram of corn stover (the residual after the corn is harvested). The table illustrates both the detail and the diversity of information available in a typical life-cycle inventory.

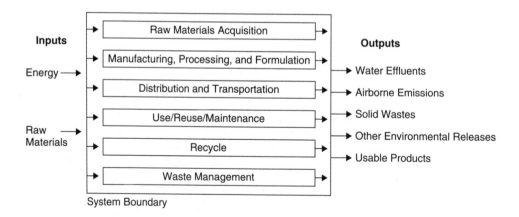

Figure 2-3 Life-cycle inventories account for materials use, energy use, wastes, emissions, and co-products over all of the stages of a product's life cycle.

Table 2-4 shows similar information for crude oil production (U.S. average data) from the same database. The entries are equally detailed but account for very different types of inputs and releases to the environment. Comparing Tables 2-3 and 2-4 can be done for some types of inputs and releases but is difficult for many others.

Table 2-3   Life-Cycle Inventory Data for the Production of 1 kg of Corn and 1 kg of Corn Stover

Inputs	Units	Value
Agrochemicals, at plant	kg	2.879E-04
Diesel, combusted in industrial equipment	L	6.860E-03
Electricity, at grid, U.S.	kWh	1.222E-02
Gasoline, combusted in equipment	L	1.881E-03
Quicklime, at plant	kg	3.050E-02
Liquefied petroleum gas, combusted in industrial boiler	L	4.873E-03
Natural gas, combusted in industrial boiler	m^3	3.051E-03
Nitrogen fertilizer, production mix, at plant	kg	1.688E-02
Phosphorous fertilizer (TSP as P_2O_5), at plant	kg	5.838E-03
Potash fertilizer (K_2O), at plant	kg	7.182E-03
Transport, train, diesel-powered	ton-km	4.397E-02
Transport, single-unit truck, diesel-powered	ton-km	1.651E-02
Land Area		
Occupation, arable, conservation tillage	m^2	4.275E-01
Occupation, arable, conventional tillage	m^2	4.684E-01
Occupation, arable, reduced tillage	m^2	2.870E-01
Water Use		
Water, river	L	4.797E+00
Water, well, in ground	L	2.776E+01
Outputs to the Environment		
Carbon dioxide, in air	kg	1.494E+00
2,4-D	kg	1.512E-06
Acetochlor	kg	2.095E-05
Alachlor	kg	2.063E-06
Ammonia	kg	1.097E-03
Atrazine	kg	4.084E-05
Bentazone	kg	1.666E-07
Bromoxynil	kg	3.652E-07
Carbofuran	kg	3.122E-07
Carbon monoxide, unspecified origin	kg	0.000E+00
Chlorpyrifos	kg	2.403E-06
Cyanazine	kg	3.601E-07
Dicamba	kg	2.122E-06

(Continued)

Table 2-3   (*Continued*)

Inputs	Units	Value
Dimethenamid	kg	5.012E-06
Dipropylthiocarbamic acid S-ethyl ester	kg	3.432E-06
Glyphosate	kg	4.505E-06
Hydrocarbons, unspecified	kg	1.728E-04
MCPA	kg	2.821E-08
Methane, unspecified origin	kg	0.000E+00
Parathion, methyl	kg	2.532E-07
Metolachlor	kg	1.656E-05
Metribuzin	kg	7.675E-08
Nitrogen oxides	kg	2.968E-03
Dinitrogen monoxide	kg	5.260E-04
Paraquat	kg	3.352E-07
Particulates, unspecified	kg	3.454E-07
Pendimethalin	kg	1.723E-06
Permethrin	kg	1.548E-07
. . . and many more agricultural chemicals		

Source: NREL, 2011

Table 2-4   Life-Cycle Inventory Data for the Production of 1 kg of Crude Oil at the Production Site

Inputs	Units	Value
Diesel, combusted in industrial boiler	L	1.252E-03
Electricity, at grid, U.S.	kWh	3.902E-02
Gasoline, combusted in equipment	L	6.843E-04
Natural gas, combusted in industrial boiler	m^3	3.278E-02
Residual fuel oil, combusted in industrial boiler	L	8.345E-04
Disposal, solid waste, unspecified, to sanitary landfill	kg	2.610E-02
Outputs to the Environment		
Methane	kg	3.530E-03
2-Hexanone	kg	2.321E-08
Acetone	kg	3.555E-08
Aluminum	kg	3.187E-04
Antimony	kg	1.992E-07
Arsenic, ion	kg	9.834E-07
Barium	kg	4.364E-03
Benzene	kg	5.963E-06
Benzene, 1-methyl-4-(1-methylethyl)-	kg	3.552E-10
Benzene, ethyl-	kg	3.355E-07
Benzene, pentamethyl-	kg	2.664E-10

Table 2-4   (*Continued*)

Inputs	Units	Value
Benzenes, alkylated, unspecified	kg	1.747E-07
Benzoic acid	kg	3.606E-06
Beryllium	kg	5.516E-08
Biphenyl	kg	1.131E-08
BOD5, biological oxygen demand	kg	6.190E-04
Boron	kg	1.116E-05
Bromide	kg	7.616E-04
Cadmium, ion	kg	1.452E-07
Calcium, ion	kg	1.142E-02
Chloride	kg	1.284E-01
Chromium	kg	8.504E-06
Cobalt	kg	7.875E-08
COD, chemical oxygen demand	kg	1.023E-03
Copper, ion	kg	1.023E-06
Cyanide	kg	2.566E-10
Decane	kg	1.036E-07
Dibenzofuran	kg	6.759E-10
Dibenzothiophene	kg	3.492E-11
Dibenzothiophene	kg	5.476E-10
Dissolved solids	kg	1.583E-01
Docosane	kg	3.804E-09
Dodecane	kg	1.966E-07
Eicosane	kg	5.413E-08
Fluorene, 1-methyl-	kg	4.046E-10
Fluorenes, alkylated, unspecified	kg	1.013E-08
Fluorine	kg	4.985E-09
Hexadecane	kg	2.146E-07
...and many more organic and inorganic chemicals		

Source: NREL, 2011

Step 3. The output from a life-cycle inventory is an extensive compilation of specific materials used and emitted, as illustrated in Tables 2-3 and 2-4. Converting these inventory elements into an assessment of environmental performance requires that the emissions and materials use be transformed into estimates of environmental impacts. Thus, the third step in a life-cycle assessment is to assess the environmental impacts of the inputs and outputs compiled in the inventory. This step is called a *life-cycle impact assessment*.

This step often involves multiple types of impacts, just as the inventory involved multiple inputs and emissions. The impacts can include natural resource use, such as energy use and water use. They can include environmental impacts, such as acid deposition, smog formation, and solid waste generation. Sometimes social impacts, such as employment, are also quantified. While calculating impacts can sometimes facilitate comparisons between disparate products or materials, often the interpretation can be challenging, as illustrated in Example 2-3.

Step 4. The fourth step in a life-cycle assessment is to interpret the results of the impact assessment, suggesting improvements whenever possible. When life-cycle assessments are conducted to compare products, for example, this step might consist of recommending the most environmentally desirable product. Alternatively, if a single product were analyzed, specific design modifications that could improve environmental performance might be suggested. This step is called an *improvement analysis* or an *interpretation step*. While comparisons between disparate products or materials can sometimes be done effectively, often the interpretation can be challenging, as illustrated in Example 2-3.

Example 2-3   Comparing cloth and disposable diapers (from Allen et al., 1992)

Disposable diapers, manufactured from paper and petroleum products, are one of the most convenient diapering systems available, while cloth diapers are often believed to be the most environmentally sound. The evidence is not so clear-cut, however. This example will quantitatively examine the relative energy and water requirements and the rates of waste generation associated with diapering systems.

Three types of diapering systems are considered in this problem: home-laundered cloth diapers, commercially laundered cloth diapers, and disposable diapers containing a superabsorbent gel. The results of life-cycle inventories for the three systems are given in the following table.

Energy Requirements and Waste Inventory per 1000 Diapers

Impact	Disposable Diapers	Commercially Laundered Cloth Diapers	Home-Laundered Cloth Diapers
Energy requirements (million BTU)	3.4	2.1	3.8
Solid waste (ft^3)	17	2.3	2.3
Atmospheric emissions (lb)	8.3	4.5	9.6
Waterborne wastes (lb)	1.5	5.8	6.1
Water requirements (gal)	1300	3400	2700

a. The authors of the report from which the data in the table are taken found that an average of 68 cloth diapers were used per week per baby. Disposable diaper usage is expected to be less because disposable diapers are changed less frequently and never require double or triple diapering. In order to compare the diapering systems, determine the number of disposable diapers required to match the performance of 68 cloth diapers, assuming the following:

   15.8 billion disposable diapers are sold annually.
   3,787,000 babies are born each year.

Children wear diapers for the first 30 months.

Disposable diapers are used on 85% of children.

b. Complete the table below. Remember to use the equivalency factor for cloth and disposable diapers determined in Part a. Based on the assumptions you made in Part a, how accurate are the entries in the table?

Ratio of Impact to Home-Laundered Impact

Impact	Disposable Diapers	Commercially Laundered Cloth Diapers	Home-Laundered Cloth Diapers
Energy requirements (million BTU)	0.50	0.55	1.0
Solid waste (ft³)			1.0
Atmospheric emissions (lb)			1.0
Waterborne wastes (lb)			1.0
Water requirements (gal)			1.0

Solution:

a. Diapers per baby per week for disposable diapers—equivalency of diapers

15.8 billion disposable diapers are sold annually.

3,787,000 babies are born each year.

Children wear diapers for the first 30 months.

Disposable diapers are used on 85% of children.

Number of babies in diapers

= (3,787,000 babies born/yr) (30 mo in diapers/12 mo/yr) = 9,467,500

Number of babies in disposable diapers = 9,467,500 babies(0.85) = 8,047,375

Number of disposable diapers per baby per year

= (15.8 × 10⁹ disposable diapers)/(8,047,375 babies) = 1963.4 disposable diapers/baby

Number of disposable diapers per baby per week

= (1963.4 disposable diapers/baby)/52 weeks = 39.3

Equivalence = (39.3 disposable diapers/baby/wk)/(68 cloth diapers/baby/wk) = 0.577

b. Complete the table "Ratio of Impact to Home-Laundered Impact"

Impact	Disposable Diapers	Commercially Laundered Cloth Diapers	Home-Laundered Cloth Diapers
Energy requirements (million BTU)	0.52	0.55	1.0
Solid waste (ft³)	4.26	1.0	1.0
Atmospheric emissions (lb)	0.50	0.47	1.0
Waterborne wastes (lb)	0.14	0.95	1.0
Water requirements (gal)	0.28	1.26	1.0

The disposable diapers create more solid waste but require less water and generate less waterborne waste than cloth diapers. Which is preferable, from the standpoint of environmental footprint, depends on the relative importance of water use and solid waste generation. This may vary from region to region and person to person. For example, in New York City, where landfill space is scarce, reducing solid waste may be more important than reducing water use. In Phoenix or Las Vegas, where water is scarce, the situation may be just the opposite.

2.3.3 Life-Cycle-Based Environmental Law

Life-cycle concepts are increasingly becoming a part of environmental regulations, especially those involving greenhouse gas emissions. For example, as this text is being written, a variety of governmental agencies are developing approaches for regulating the emissions of greenhouse gases associated with the production and use of transportation fuels. As an example, the Energy Independence and Security Act (HR 6) of 2007 states (EISA, 2007):

> No Federal agency shall enter into a contract for procurement of an alternative or synthetic fuel, including a fuel produced from nonconventional petroleum sources, for any mobility-related use, other than for research or testing, unless the contract specifies that the lifecycle greenhouse gas emissions associated with the production and combustion of the fuel supplied under the contract must, on an ongoing basis, be less than or equal to such emissions from the equivalent conventional fuel produced from conventional petroleum sources.

States are also implementing greenhouse gas emission regulations. For example, the California Global Warming Solutions Act of 2006 (2006) has resulted in draft regulations that establish a limit for life-cycle greenhouse gas emissions of transportation fuels. Both the California low-carbon fuel standard and Section 526 of EISA, cited previously, require a life-cycle evaluation of the greenhouse gas emissions of transportation fuels, and this is becoming a common approach to considering greenhouse gas emissions. Employing a life-cycle approach in estimating greenhouse gas emissions from the production and use of transportation fuels means assessing all emissions from field to vehicle tank and from tank to vehicle exhaust. This scope of emissions assessment is frequently referred to as a "well-to-wheels" analysis.

The steps involved in performing a life-cycle assessment (scope and boundaries definition, life-cycle inventories, life-cycle impact assessment, life-cycle interpretation) can be applied to the problem of compliance with regulations such as EISA. The U.S. Air Force, the largest consumer of fuel in the federal government, must comply with Section 526 of EISA in purchasing fuels. To provide guidance for producers of fuel seeking to sell Section 526—compliant fuels, the Air Force convened a working group to develop guidance on procedures for estimating life-cycle greenhouse gas emissions for aviation fuels. This group defined the following steps for an LCA in this application (adapted from Allen et al., 2009):

Step 1: Determine the goal and scope of the assessment. Choices made at the goal- and scope-setting stage can significantly impact the results of a life-cycle assessment. For example, an LCA could be based on the operation of a specific refinery, or the average operations of all refineries in a state, region, or nation. Also, the analysis could include not only industrial- and combustion-related greenhouse gas emissions but also the implications of changes in land use at local, regional, national, or global scales.

Step 2: Develop an inventory of the greenhouse gas emissions throughout the life-cycle system. Multiple choices made at this stage also have the potential to significantly impact the results of the analysis. Time periods and spatial scales for data gathering and strategies for filling data gaps are among multiple factors influencing the results of the analyses. For example, EISA Title II, Subtitle A, Section 201, requires the greenhouse gas emissions of alternative fuels to be compared to the greenhouse gas emissions of average petroleum-based fuels in 2005. In 2005, disruptions due to Hurricanes Katrina and Rita had substantial impacts on refining operations. Other years without these disruptions may have different greenhouse gas emission characteristics, highlighting the importance of the selection of time periods for analysis. Furthermore, choices made at this stage regarding how emissions are assigned to processes that produce multiple products have a significant impact. For example, soy grown to make soy oil for biodiesel also results in the production of soy meal. How are the emissions associated with growing the soybeans distributed among the soy oil and the soy meal products? By mass? By economic value? By some other measure? The choice can substantially influence the results of the analysis (Allen et al., 2009).

Step 3: Assess the climate change impacts of the life-cycle inventory. Although methods for assessing global warming potentials (GWPs) of emissions are available from the IPCC (IPCC 2007), choices made at this stage still influence results of life-cycle assessments. For example, many LCAs consider only high-volume emissions (e.g., emissions of CO_2, CH_4, and N_2O), omitting other emissions that influence the energy balance of the atmosphere (such as sootlike "black carbon").

Step 4: Interpret the LCA results. Interpretation of LCA results needs to consider the types of uncertainties outlined in steps 1 through 3. A detailed case study of a life-cycle assessment of a fuel that could be used to determine EISA compliance is provided by Allen et al. (2011).

This example of the application of life-cycle approaches illustrates that, while life-cycle frameworks for analyzing environmental issues can be extremely useful, there is a need to develop detailed guidance for the implementation of the frameworks. Risk-based frameworks for assessing and managing environmental risks evolved over decades. Life-cycle frameworks will similarly need to evolve.

2.4   LIFE-CYCLE ASSESSMENT TOOLS

A number of quantitative tools are emerging that enable life-cycle assessments and analyses. These tools fall into two general types: process-based analysis tools and input-output analysis tools. This section will examine the types of analyses enabled by both of these types of tools, using tools that are in the public domain.

2.4.1   Process-Based Life-Cycle Assessments

Process-based life-cycle assessments follow supply chains, as illustrated in Figure 2-1. For example, for a biofuel, a process-based life-cycle assessment would link together process steps including corn growing, transporting the corn to a refiner, refining cornstarch or corn stover into a fuel, transporting the fuel to the point of sale, and combusting the fuel. As shown in Table 2-3, life-cycle inventory data can be assembled for each of these steps.

The steps are linked by accounting for mass flows. For example, assume that 500 kg of corn and 500 kg of corn stover were grown. The initial inputs and emissions would be equal to 500 times the amounts listed in Table 2-3. If the harvested material now had to be transported to a refiner 100 km from the field where the corn plant was grown, 1 metric ton of material would need to be transported 100 km for a total transportation burden of 100 ton-km (tkm). If the fuel requirement for transporting 1 tkm of freight by truck is 0.027 L of diesel fuel (NREL, 2011), then the 2.7 L of diesel would be added to the diesel requirements of the biofuel produced from 500 kg of corn and 500 kg of stover. Similar additions would be made to all of the inputs and emissions. Then, the next process step would be added to the analysis, and this process would continue until the entire life cycle was modeled.

Tools exist in the form of data on individual process steps, and in the form of software packages that facilitate the linking of individual processes. An example of the former category is the U.S. Life Cycle Inventory (LCI) Database, maintained by the National Renewable Energy Laboratory (NREL, 2011). An example of the latter category, which is in the public domain (there are multiple software packages that can be licensed), is the GREET model. GREET, the Greenhouse Gases, Regulated Emissions, and Energy Use in Transportation model, is maintained by Argonne National Laboratory (GREET, 2011).

There are several major challenges associated with performing process life-cycle assessments. One challenge is availability of data. A review of the U.S. LCI database (NREL, 2011) reveals that while data are available for many commodity materials, there are many data gaps. A second issue is that of system boundaries. For example, the data for corn growing includes the fuel used for the tractor that plowed the fields. But what about the steel used to make the tractor? What about the materials used to construct the steel mill that made the steel that went into the tractor that plowed the fields? Where do we draw the boundary? The answers are not simple, but recently another form of life-cycle assessment that does not have these challenges has emerged: input-output LCA.

2.4.2 Input-Output LCA

An input-output LCA relies on tools that have become widely used by economists. These economic input-output tools segment national and regional economies into sectors and follow the flows of money. Consider a simple example. Imagine that a consumer purchases an automobile for $20,000. The automaker might spend $10,000 purchasing parts from first-tier suppliers. Those first-tier suppliers then might spend $1000 on steel. The steelmaker in turn might purchase coal. The transactions would continue, with the initial purchase leading to more than $50,000 eventually changing hands. Economists have built models that define these financial linkages between sectors of the economy. Typically, an economy is broken into hundreds of sectors and the financial flows between each sector and all of the other sectors are tracked, creating an economic input-output model (EIO). The EIO can be used as a life-cycle assessment tool by recognizing that for each sector, parameters such as energy use per dollar of sales can be tracked. If dollar flows between sectors are known, and if energy use per dollar is known for each sector, energy use across the economy can be tracked. This approach to modeling economy-wide flows of energy, materials, and emissions is known as an EIO-LCA. EIO-LCA models are relatively recent developments, but online tools are available (e.g., see Problem 8 at the end of the chapter).

The advantage of EIO-LCA approaches is that they track all flows up to the point of purchase. Thus, they avoid problems of system boundaries. Only limited types of flows are tracked (e.g., energy, greenhouse gases, certain toxic compounds), so this method also suffers from data gaps. It also has the disadvantage of representing products at a relatively coarse level. So, for example, all automobiles, including electric vehicles, small sedans, and large luxury cars, are averaged in the same economic sector.

2.4.3 Hybrid Approaches

Process-based and EIO-LCAs have complementary strengths and are beginning to be used in sophisticated ways. It is beyond the scope of this chapter to describe these emerging, advanced tools, but the problems at the end of this chapter will provide an introduction to the types of analyses that can be done using various analysis tools.

2.5 SUMMARY

Complex environmental and sustainability issues are best managed through structured analysis frameworks. This chapter has provided summaries of both traditional (risk-based) and emerging (life-cycle-based) frameworks. The basic principles, methodologies, and applications in environmental regulation have been summarized for both methods.

PROBLEMS

1. **Voluntary risk** Each year, approximately 45,000 persons lose their lives in automobile accidents in the United States (population 281 million according to the 2000 census). How many fatalities would be expected over a three-day weekend in the Minneapolis–St. Paul, Minnesota, metropolitan area (population 2 million)?

2. **Involuntary risk** Lurmann et al. (1999) have estimated the costs associated with ozone and fine particulate matter concentrations above the NAAQS in Houston. They estimated that the economic impacts of early mortality and morbidity associated with elevated fine particulate matter concentrations (above the NAAQS) are

approximately $3 billion/year. Hall et al. (1992) performed a similar assessment for Los Angeles. In the Houston study, Lurmann et al. examined the exposures and health costs associated with a variety of emission scenarios. One set of calculations demonstrated that a decrease of approximately 300 tons/day of fine particulate matter emissions resulted in a 7 million person-day decrease in exposure to particulate matter concentrations above the proposed NAAQS for fine particulate matter, 17 fewer early deaths per year, and 24 fewer cases of chronic bronchitis per year. Using estimated costs of $300,000 per case of chronic bronchitis and $7,000,000 per early death, estimate the social cost per ton of fine particulate matter emitted.

3. **Life cycles of cups** In evaluating the energy implications of the choice between reusable and single-use cups, the energy required to heat wash water is a key parameter. Consider a comparison of single-use polypropylene (PP) and reusable PP cups. The reusable cup has a mass roughly 14 times that of the single-use cup (45 g versus 3.2 g), which, in turn requires petroleum feedstocks.

   a. Calculate the number of times the reusable cup must be used in order to recoup the energy in the petroleum required to make the reusable cup.

   b. Assuming that the reusable cup is washed after each use in 0.27 L of water, and that the wash water is at 80°C (heated from 20°C), calculate the energy used in each wash if the water is heated in a gas water heater with an 80% efficiency. Calculate the number of times the reusable cup must be used in order to recoup both the energy required to make the reusable cup and the energy used to heat the wash water. Assume that 1.2 kg of petroleum are required to produce 1 kg of polypropylene and that the energy of combustion of petroleum is 44 MJ/kg.

   c. Repeat Part b, assuming that an electric water heater is used (80% efficiency) and that electricity is generated from fuel at 33% efficiency.

$$C_p \text{ of water} = 4.184 \text{ J/g K}$$

4. **Durability versus efficiency improvements in newer products** In minimizing the environmental footprints of products, there is tension between product durability and rapidly replacing older products with newer products that have less environmental impact associated with their use. Consider this question: When is it most energy-efficient to replace my vehicle?

   a. The production of a 1995 vehicle consumed 125,000 MJ of energy, and the energy intensity of the materials used in manufacturing automobiles (energy required per kilogram of material) decreases by 1% to 2% per year. Assuming that the energy intensity of automobile manufacturing decreased by 1.5% per year between 1990 and 2010, calculate the energy required to produce a new automobile during the model years 1990, 2000, 2005, and 2010.

   b. The projected average fuel economy of light-duty automobiles is expected to increase from 27.5 to 32.5 mpg between 1990 and 2020. Assume that this increase occurs in step changes, with an average fuel economy of 27.5 mpg between 1990 and 1999, 30 mpg between 2000 and 2009, and 32.5 mpg between 2010 and 2019. Calculate the amount of energy used (assuming an energy content for gasoline of 124,000 BTU/gal, $1.3 * 10^8$ J/gal) for vehicles traveling 12,000 miles per year for the decades of the 1990s, 2000s, and 2010s.

   c. Is it more efficient to replace a vehicle every 10 years or every 15 years?

5. **Options for moving energy** Approximately 1 billion tons of coal are burned annually in the United States, providing 50% of the country's electricity consumption. The coal may be either moved by train from the mine to power plants near where the power is used, or combusted near the mine mouth to generate electricity that can be transmitted over long distances to the users. As a case study of this trade-off, consider electricity use in Dallas, which is generated, in part, using coal from the Powder River Basin (PRB) in Wyoming. Power plants using PRB coal supply 6.5 billion kWh of power per year to the Dallas area, at an average conversion efficiency (energy in the generated electricity per energy in the fuel burned) of 33%. The coal mined at the PRB has a heat content of 8340 BTU/lb coal (1 kWhr = 3412 BTU).

   a. Determine the amount of coal required from the PRB to support consumers in Dallas.

   b. If the energy required to transport coal by train is 0.0025 gallons of diesel per ton mile, and the distance from the PRB to Dallas is 1000 miles, calculate the amount of energy required to transport the coal to Texas, and the total energy consumed in combustion and transport. What fraction of the total energy consumption is due to transport? Assume that diesel fuel has an energy content of 124,000 BTU/gal.

   c. Calculate the amount of coal consumed if the electricity were generated at the mine (assume a 33% power plant efficiency) and if the transmission losses for the electricity, from the mine to Dallas, were 7%.

   d. Which option (transporting coal or transporting electricity) would be more efficient?

6. **Functional unit in life-cycle assessment: personal mobility** Mobility is one of the measures of quality of life that citizens of many developed nations value highly, ranked behind only food and shelter as necessities for life. Mobility is also a key factor in sustainability because of the cumulative effects of providing mobility on the environment, on resource depletion, and on the economy.

   In the table below, data are presented on two modes of transportation, automobile and bus. Use these data to answer the questions that follow.

Annual Average Personal Transportation Data for the United States

Automobiles (cars)	27.5 mpg gasoline (2010 Corporate Average Fuel Economy [CAFE] std.) 1.6 persons per automobile
Buses	mpg diesel (est.) 30 persons per bus (est.)

Source: www1.eere.energy.gov/vehiclesandfuels/facts/2010_fotw613.html

Other data and conversion factors: 150,000 BTU/gal gasoline, 163,000 BTU/gal diesel (includes production energy and feedstock energy over the fuel life cycle):

$$4.3 + 19.4 \text{ lb } CO_2 \text{ e/gal gasoline (production + combust.)}$$
$$3.6 + 22.2 \text{ lb } CO_2 \text{ e/gal diesel (production + combust.)}$$

   a. Define an appropriate functional unit for a comparison of the bus and car transportation table for personal mobility.

   b. Calculate the gallons of fuel needed to satisfy the transportation functional unit, and then convert gallons to energy (BTU per functional unit). Also, calculate the CO_2 emissions per functional unit (pounds of CO_2 emitted per functional unit).

    c. Compare bus and auto transport based on energy consumption and greenhouse gas emissions.

7. Functional unit in life-cycle assessment: transport of goods Transport of goods is another important energy-consuming and greenhouse-gas-emitting activity, and, as for personal mobility, there are choices in modes of freight transportation.

    In the table below, data are presented on three modes of freight transportation: by road (heavy trucks), by rail, and by ship (oceanic freighter). Use these data to answer the questions that follow.

Annual Average Freight Transportation Data for the United States

Road (heavy truck)	1 gallon diesel transports 20 tons 5.5 miles
Rail	1 gallon diesel transports 1 ton 423 miles
Ship (oceanic freighter)	1 gallon heavy oil transports 1 ton 1500 miles

Source: Transportation Energy Data Book, U.S. Department of Energy, 2010

    Other data and conversion factors: 190,000 BTU/gal heavy oil, 163,000 BTU/gal diesel

$$3.7 + 26.0 \text{ lb } CO_2 \text{ e/gal heavy oil (production + combust.)}$$
$$3.6 + 22.2 \text{ lb } CO_2 \text{ e/gal diesel (production + combust.)}$$

    a. Define an appropriate functional unit for a comparison of the transportation modes shown in the table for freight transportation.

    b. Calculate the gallons of fuel needed to satisfy the freight transportation functional unit, and then convert gallons to energy (BTU per functional unit). Also, calculate the CO_2 emissions per functional unit (pounds of CO_2 emitted per functional unit).

    c. Rank the transportation modes from the least energy- and greenhouse-gas-intensive to the most energy- and greenhouse-gas-intensive.

8. Transport of goods: truck or air? Use the U.S. Life Cycle Inventory Database (www.nrel.gov/lci) to determine the relative amount of diesel fuel required to transport 1 ton of freight 1000 km by truck and by air.

9. Economic input-output life-cycle assessment Review the input-output model for life-cycle assessment, developed by Carnegie Mellon University. This model, available at the Web site www.eiolca.net, allows you to estimate the overall environmental impacts from expending a user-defined dollar amount in any of roughly 400 economic sectors in the United States. It provides rough guidance on the relative impacts of different types of products, materials, services, and industries, up to the point of purchase.

    Use the model to answer these questions:

    a. What is the most energy-intensive sector of the chemical industry (resin, rubber, artificial fibers, agricultural chemicals, and pharmaceuticals sector in the EIOLCA model; measured as total life cycle energy use per million dollars of sales in the sector)?

    b. What suppliers to the automotive manufacturing sector have the greatest emissions of greenhouse gases?

c. How much energy is used to manufacture a passenger vehicle costing $20,000? How does this compare to the energy consumption from driving the vehicle? Assume that the car is driven 200,000 miles and gets 30 mpg of gasoline consumed. Assume that the gasoline has a heating value of 124,000 BTU/gal and that it takes 26,000 BTU of energy to produce the gasoline.

APPENDIX: READILY AVAILABLE HAZARD REFERENCES

Although the list is not comprehensive, listed below are references commonly used to inform hazard assessment. The list is intended as a starting point for the engineer charged with hazard assessment.

1. **MSDS.** The Material Safety Data Sheet is a document developed by chemical manufacturers. The MSDS contains safety and hazard information, physical and chemical characteristics, and precautions on safe handling and use. MSDS may include hazards to animals, especially aquatic species. The manufacturer is required to keep it up-to-date. Any employer that purchases a chemical is required by law to make the MSDS available to employees. Development of an MSDS is required under OSHA's Hazard Communication Standard.

2. *NIOSH Pocket Guide to Chemical Hazards.* NIOSH is the National Institute for Occupational Safety and Health; this is the organization that performs research for OSHA, the Occupational Safety and Health Administration. The *Pocket Guide* may be found online at www.cdc.gov/niosh/npg/. It includes safety information, some chemical properties, and OSHA Permissible Exposure Limit concentrations, or PELs. The lower the permissible concentration, the greater the hazard to human health.

3. **IRIS.** IRIS is a database maintained by the U.S. Environmental Protection Agency. IRIS stands for Integrated Risk Information System. It is available through www.epa.gov/ngispgm3/iris/index.html. IRIS is a database of human health effects that may result from exposure to various substances found in the environment.

4. **Hazardous Substances Data Bank (HSDB).** The data bank is available from the National Library of Medicine. The Web address is http://toxnet.nlm.nih.gov. The HSDB is a toxicology data file that focuses on the toxicology of potentially hazardous chemicals. It is enhanced with information on human exposure, industrial hygiene, emergency handling procedures, environmental fate, regulatory requirements, and related areas.

5. **Toxnet.** Toxnet is also available from the National Library of Medicine. The Web address is http://toxnet.nlm.nih.gov. Toxnet is a cluster of databases on toxicology, hazardous chemicals, and related areas. Both IRIS and the HSDB are available through Toxnet.

6. **Casarett and Doull's text** *Toxicology: The Basic Science of Poisons, Fifth Edition* **(1996).** This is the classic text in the field for interested readers. It is published by McGraw-Hill.

7. *Patty's Industrial Hygiene and Toxicology.* This set of volumes is a starting point for readers who want more information than exposure limits but who are not experts in toxicology. It is published by John Wiley & Sons.

REFERENCES

Allen, D. T., and D. R. Shonnard. 2002. *Green Engineering: Environmentally Conscious Design of Chemical Processes.* Upper Saddle River, NJ: Prentice Hall.

Allen, D. T., N. Bakshani, and K. S. Rosselot. 1992. *Pollution Prevention: Homework and Design Problems for Engineering Curricula.* New York: American Institute of Chemical Engineers.

Allen, D. T., C. Allport, K. Atkins, J. S. Cooper, R. M. Dilmore, L. C. Draucker, K. E. Eickmann, J. C. Gillen, W. Gillette, W. M. Griffin, W. E. Harrison III, J. I. Hileman, J. R. Ingham, F. A. Kimler III, A. Levy, C. F. Murphy, M. J. O'Donnell, D. Pamplin, G. Schivley, T. J. Skone, S. M. Strank, R. W. Stratton, P. H. Taylor, V. M. Thomas, M. Wang, and T. Zidow. 2009. *The Aviation Fuel Life Cycle Assessment Working Group, Framework and Guidance for Estimating Greenhouse Gas Footprints of Aviation Fuels (Final Report).* Prepared for Universal Technology Corporation and the Air Force Research Laboratory. April Interim Report. AFRL-RZ-WP-TR-2009-2206.

Allen, D. T., C. Allport, K. Atkins, D. G. Choi, J. S. Cooper, R. M. Dilmore, L. C. Draucker, K. E. Eickmann, J. C. Gillen, W. Gillette, W. M. Griffin, W. E. Harrison III, J. I. Hileman, J. R. Ingham, F. A. Kimler III, A. Levy, J. Miller, C. F. Murphy, M. J. O'Donnell, D. Pamplin, K. Rosselot, G. Schivley, T. J. Skone, S. M. Strank, R. W. Stratton, P. H. Taylor, V. M. Thomas, M. Q. Wang, and T. Zidow. 2011. *Life Cycle Greenhouse Gas Analysis of Advanced Jet Propulsion Fuels: Fischer-Tropsch Based SPK-1 Case Study.* Final Report from the Aviation Fuel Life Cycle Assessment Working Group to the U.S. Air Force. AFRL-RZ-WP-TR-2010-XXXX. Draft, February 4.

California Global Warming Solutions Act of 2006. 2006. Assembly Bill 32 (AB 32). Available at www.leginfo.ca.gov/pub/05-06/bill/asm/ab_0001-0050/ab_32_bill_20060927_chaptered.pdf. Accessed March 2011.

Consoli, F., D. T. Allen, I. Boustead, J. Fava, W. Franklin, A. A. Jensen, N. deOude, R. Parrish, R. Perriman, D. Postlewaite, B. Quay, J. Sequin, and B. Vigon. 1993. *Guidelines for Life Cycle Assessment: A Code of Practice.* Pensacola, FL: SETAC Press.

EISA (Energy Independence and Security Act of 2007). 2007. Available at http://frwebgate. access.gpo.gov/cgi-bin/getdoc.cgi?dbname=110_cong_bills&docid=f:h6enr.txt.pdf. Accessed March 2011.

Fan, A. M., and L. W. Chang, eds. 1996. *Toxicology and Risk Assessment: Principles, Methods, and Applications.* New York: Marcel Dekker, Inc., p. 247.

Federal Focus, Inc. 1991. *Towards Common Measures: Recommendations for a Presidential Executive Order on Environmental Risk Assessment and Risk Management Policy.* Washington, DC: Federal Focus, Inc., and the Institute for Regulatory Policy.

Fort, D. D. 1996. "Environmental Laws and Risk Assessment." In *Toxicology and Risk Assessment: Principles, Methods, and Applications*, edited by A. M. Fan and L. W. Chang. New York: Marcel Dekker, Inc., pp. 653–77.

GREET (Greenhouse Gases, Regulated Emissions, and Energy Use in Transportation Model). 2011. Argonne National Laboratory. Available at http://greet.es.anl.gov/. Accessed July 2011.

Hall, J. V., A. M. Winer, M. T. Kleinman, F. W. Lurmann, V. Brajer, and S. D. Colome. 1992. "Valuing the Health Benefits of Clean Air." *Science* 255:812–17.

IPCC (Intergovernmental Panel on Climate Change). 2007. *Climate Change 2007: The Physical Science Basis*. Contribution of Working Group I to the Fourth Assessment Report of the IPCC. Cambridge: Cambridge University Press.

ISO (International Standards Organization). 2006. 14040 series of standards. Available at www.iso.org/iso/iso_14000_essentials.

Lurmann, F. W., J. V. Hall, M. Kleinman, L. R. Chinkin, V. Brajer, D. Meacher, F. Mummery, R. L. Arndt, T. H. Funk, S. H. Alcorn, and N. Kumar. 1999. *Assessment of the Health Benefits of Improving Air Quality in Houston, Texas*. Final Report by Sonoma Technologies to the City of Houston. STI-998460-1875-FR. November.

NRC (National Research Council). 1983. *Risk Assessment in the Federal Government: Managing the Process*. Committee on Institutional Means for Assessment of Risks to Public Health. Washington, DC: National Academy Press.

————. 2009. *Science and Decisions: Advancing Risk Assessment*. Washington, DC: National Academy Press.

NREL (National Renewable Energy Laboratory). 2011. U.S. National Life Cycle Inventory Database. Available at www.nrel.gov/lci. Accessed July 2011.

Presidential/Congressional Commission on Risk Assessment and Risk Management. 1997. *Final Report, Vol. 1*, p. 1.

Roberts, W. C., and C. O. Abernathy. 1996. "Risk Assessment: Principles and Methodologies." In *Toxicology and Risk Assessment: Principles, Methods, and Applications*, edited by A. M. Fan and L. W. Chang. New York: Marcel Dekker, Inc., pp. 245–70.

U.S. EPA (U.S. Environmental Protection Agency). 2011a. *Clean Air Act, Section 812 Analyses on the Benefits and Costs of the Clean Air Act from 1990 to 2020*. March. Available at www.epa.gov/oar/sect812.

————. 2011b. *The Benefits and Costs of the Clean Air Act from 1990 to 2020*. March. Available at www.epa.gov/oar/sect812/feb11/summaryreport.pdf and www.epa.gov/oar/sect812/feb11/fullreport.pdf.

CHAPTER 3

Environmental Law
and Regulation

3.1   INTRODUCTION

Engineers practice a profession and are required to obey specific laws governing their professional conduct. One important class of laws that all engineers should be aware of is environmental law. These laws are designed to protect human health, natural resources, and the environment by placing limits on the quantity, chemical makeup, and methods of disposal of environmental releases and wastes. Some of these laws restrict releases into the air and water; some place restrictions on the manner in which hazardous waste is stored, transported, treated, and disposed. Other laws place strict liability on the generators of hazardous waste, requiring responsible parties to clean up sites that become contaminated. For manufacturers of new substances, there are regulatory requirements that need to be adhered to before introducing a new substance into the marketplace. Also emerging are international agreements that seek to preserve the sustainability of global resources.

   The purpose of this chapter is to provide an overview of environmental regulations and the types of global agreements that are emerging around the issue of sustainability. Much of the material on environmental regulations in this chapter has been adapted from the review of environmental law by Lynch (1995) and Chapter 3 in the text *Green Engineering* (Allen and Shonnard, 2002). More comprehensive sources on this topic include the United States Code (U.S.C.), which are the federal laws regulating practices that may impact public health and the environment, and the Code of Federal Regulations (C.F.R.), which are the environmental regulations that implement the federal laws. *The Environmental Law Handbook* (Sullivan and Adams, 1997) and West's *Environmental Law Statutes* (West Law School, 2011) are both compendia of existing statutes. Most of these sources can be found online.

There are approximately 20 major federal statutes, hundreds of state and local ordinances, thousands of federal and state regulations, and even more federal and state court cases and administrative adjudications that deal with environmental issues. Taken together, they make up the field of environmental law, which has seen explosive growth in the last 40 years, as shown in Figure 3-1. *Environmental regulations* and the *common law system* of environmental law regulate behavior by affected entities. For example, the Clean Water Act (an environmental statute) requires facilities that discharge pollutants from a point source into navigable waters of the United States to apply for a National Pollutant Discharge Elimination System (NPDES) permit. In many firms, engineers are responsible for preparing the application to obtain these permits. The common law (law created by court decisions) encourages engineers to act responsibly when performing their professional duties because environmental laws and regulations do not cover all instances in which actions of individuals and organizations may harm public health or the environment. In addition, engineers need to be aware of environmental laws and regulations in order to protect themselves and their organizations from restrictions, fines, and legal action.

The sources of environmental laws and regulations are *legislatures, administrative agencies*, and the *courts*. When drafting environmental laws, federal and state

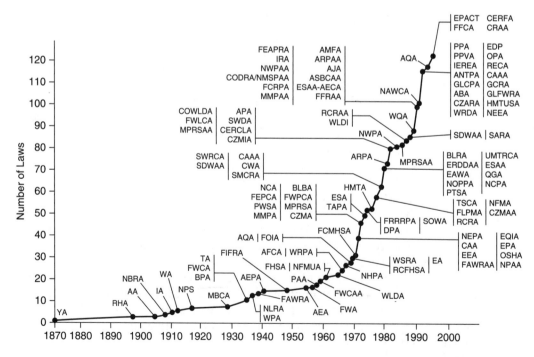

Figure 3-1   Cumulative growth in federal environmental laws and amendments (Allen and Shonnard, 2001)

*Statutes
Ordinances
regulation
Cases*

legislatures often use vague language regarding specific regulatory requirements, discharge limits, and enforcement provisions. Often, legislators do not have the time or expertise to determine how the statute is to be implemented and therefore leave the detailed development of regulations to administrative agencies.

Administrative agencies, such as the EPA, give meaning to statutory provisions through a procedure known as *rule making*. Federal rule making consists of publishing proposed regulations in the *Federal Register*, providing an opportunity for public comment, and then publishing final regulations in the *Federal Register*, which have the force of law. As such, administrative agencies fulfill a legislative function. Administrative agencies can be created as part of the executive, legislative, or judicial branches of government. In 1970, President Richard Nixon established the EPA in order to consolidate federal programs for controlling air and water pollution, radiation, pesticides, and solid waste disposal. Administrative agencies are created by statute (for example, the Occupational Safety and Health Act established the Occupational Safety and Health Administration), and the agency powers are derived from their enabling legislation. Administrative agencies also have the authority to resolve disputes that arise from the exercise of their administrative powers. Regulated entities have the right to appeal decisions made by administrative agencies to an administrative law court created by the statute. Thus administrative agencies have a judicial function in addition to a legislative function.

The courts are the third government actor that helps to define the field of environmental law. The roles of the courts in environmental law are

1. To determine the coverage of environmental statutes (which entities are subject to the regulations)
2. To review administrative rules and decisions (ensuring that regulations are properly promulgated and within the statutory authority granted to the agency)
3. To develop the common law system (a record of individual court cases and decisions that set a precedent for future judicial actions)

An example of court action in the area of environmental law is the 2007 Supreme Court decision in the case of *Massachusetts v. EPA*. By a vote of 5 to 4, the Supreme Court ruled that the U.S. Environmental Protection Agency has the statutory authority to regulate CO_2 and other greenhouse gases under the Clean Air Act. Other examples of court actions include imposing civil and criminal penalties for violations of statutes.

The goal of Section 3.2 is to provide a brief description of the most important features of nine federal environmental statutes that significantly affect engineers. This brief survey is meant to be representative, not comprehensive, and the focus will be on federal laws because they have national scope and often serve as models for state environmental statutes. We will begin with three statutes that regulate the creation, use, and manufacture of chemical substances. Next, we will cover the key provisions of three statutes that seek to control the discharge of pollutants to specific environmental media: to the air, water, and soil. Next, a statute that initiated a cleanup program for the many sites of soil and groundwater contamination will be discussed. The final two statutes involve the reporting of toxic substance releases and a voluntary program for preventing pollution generation and release at industrial facilities. Section 3.3 will describe the evolution in environmental regulation from end-of-pipe pollution control to more proactive pollution prevention and sustainability approaches. Section 3.4 presents the key features of pollution prevention, including its position in the hierarchy of environmental management alternatives; a short review of terminology used in the practice of pollution prevention; and examples of pollution prevention strategies and applications.

3.2   NINE PROMINENT FEDERAL ENVIRONMENTAL STATUTES

This section provides the key provisions of nine federal environmental statutes. Taken together, these laws regulate materials and products throughout their life cycle, from creation and production, to use and disposal. The nine laws that will be outlined are

1. The Toxic Substances Control Act (TSCA), 1976 (the import and manufacture of new substances and the testing and restriction of substances already in commerce)
2. The Federal Insecticide, Fungicide, and Rodenticide Act (FIFRA), 1947 (the registration, labeling, and use of pesticides)
3. The Occupational Safety and Health Act (OSH Act), 1970 (the provision of safe and healthful working conditions)
4. The Clean Air Act (CAA), 1955 (air pollutant emissions)
5. The Clean Water Act (CWA), 1972 (water pollution discharge; maintenance of water quality in lakes, rivers, and streams; regulation of subsurface disposal of wastes; and provision of funding for facilities to supply drinking water and to treat wastewater)
6. The Resource Conservation and Recovery Act (RCRA), 1976 (the regulation of treatment, storage, and disposal of hazardous and non-hazardous wastes)
7. The Comprehensive Environmental Response, Compensation, and Liability Act (CERCLA), 1980 (the cleanup of abandoned and inactive hazardous waste sites)
8. The Emergency Planning and Community Right-to-Know Act (EPCRA), 1986 (responding to emergencies and reporting of toxic chemical usage)
9. The Pollution Prevention Act (PPA), 1990 (a proactive approach to reducing environmental impact)

A summary of these prominent federal environmental statutes is provided in Table 3-1. The most important regulatory provisions for each statute are stated

Table 3-1 Summary Table for Environmental Laws

Environmental Statute	Date Enacted	Background	Key Provisions
Regulation of Chemical Manufacturing			
The Toxic Substances Control Act (TSCA)	Enacted 1976 Amended 1986 1988 1990 1992	Toxic substances, such as p olychlorinated biphenyls (PCBs), began appearing in the environment and in food supplies. This prompted the federal government to create a program to assess the risks of chemicals before they are introduced into commerce.	Manufacturers, importers, and processors may be required to submit a report detailing chemical identity, uses of the substance, and numbers of persons exposed to the substance for each chemical they manufacture, import, or process. Extensive testing by companies may be required for chemicals of concern. A Premanufacture Notice (PMN) must be submitted before manufacturing and selling a chemical substance.
The Federal Insecticide, Fungicide, and Rodenticide Act (FIFRA)	Enacted 1947 Amended 1972 1988 1996	Because all pesticides are toxic to some plants and animals, they may pose an unacceptable risk to human health and the environment. FIFRA is a federal statute whose purpose is to assess the risks of pesticides and to control their usage so that any exposure that may result is within an acceptable level of risk.	Before any pesticide can be distributed or sold in the United States, it must be registered with the EPA. The registration data are difficult and expensive to develop and must prove that the chemical is effective and safe to humans and the environment. Labels must be placed on pesticide products that indicate approved uses and restrictions.
The Occupational Safety and Health Act (OSH Act)	1970	The agency that oversees the implementation of the OSH Act is the Occupational Safety and Health Administration (OSHA). All private facilities having more than ten employees must comply with the OSH Act requirements.	Companies must adhere to all OSHA health standards (exposure limits to chemicals) and safety standards (physical hazards from equipment). The OSH Act's Hazard Communication Standard requires companies to prepare a material safety data sheet (MSDS), label chemical substances, and inform and train employees in the safe use of chemicals.
Regulation of Discharges to the Air, Water, and Soil			
Clean Air Act (CAA) (originally enacted as the Air Pollution Control Act of 1955)	Enacted 1955 Amended 1963 1967 1970 1977 1990	The CAA is intended to maintain air quality by establishing uniform ambient air quality standards that are protective of human health or within the capabilities of the best available treatment technologies. The CAA also addresses specific air pollution problems such as hazardous air pollutants, stratospheric ozone depletion, acid rain, and possibly greenhouse gases.	The CAA established the National Ambient Air Quality Standards (NAAQS) for maximum concentrations in ambient air of carbon monoxide, lead, oxides of nitrogen, ozone, particulate matter, and SO_2. States must develop source-specific emission limits to achieve the NAAQS. States issue air emission permits to facilities. Stricter requirements are often established for hazardous air pollutants (HAPs) and for new sources.

(Continued)

Table 3-1 (*Continued*)

Environmental Statute	Date Enacted		Background	Key Provisions
Clean Water Act (CWA)	Enacted	1972	The CWA is the first comprehensive federal program designed to reduce pollutant discharges into the nation's waterways ("zero discharge" goal). Another goal of the CWA is to make water bodies safe for swimming, fishing, and other forms of recreation ("swimmable" goal). This act has resulted in significant improvements in the quality of the nation's waterways since its enactment.	The CWA established the National Pollutant Discharge Elimination System (NPDES) permit program that requires any point source of pollution to obtain a permit. Permits either contain effluent limits or require the installation of specific pollutant treatment. Permit holders must monitor discharges, collect data, and keep records of the pollutant levels in their effluents. Industrial sources that discharge into sewers must comply with EPA pretreatment standards by applying the best available control technology (BACT).
	Amended	1977 1987		
Resource Conservation and Recovery Act (RCRA)	Enacted	1976	The RCRA was enacted to regulate the "cradle-to-grave" generation, transport, and disposal of both non-hazardous and hazardous wastes on or in the land, encourage recycling, and promote the development of alternative energy sources based on solid waste materials.	Generators must maintain records of the quantity of hazardous waste generated and where the waste was sent for treatment, storage, or disposal and file this data in biennial reports to the EPA. Transporters and disposal facilities must adhere to similar requirements for record keeping as well as for monitoring the environment.
	Amended	1984		

Cleanup, Emergency Planning, and Pollution Prevention

Environmental Statute	Date Enacted		Background	Key Provisions
The Comprehensive Environmental Response, Compensation, and Liability Act (CERCLA)	Enacted	1980	CERCLA began a process of identifying and remediating uncontrolled hazardous waste disposal at abandoned sites, industrial complexes, and federal facilities. The EPA is responsible for creating a list of the most hazardous sites of contamination, which is termed the National Priority List (NPL). CERCLA was amended by the Superfund Amendments and Reauthorization Act (SARA) of 1986.	After a site is listed in the NPL, the EPA identifies potentially responsible parties (PRPs) and notifies them of their potential CERCLA liability, which is strict, joint and several, and retroactive. PRPs are (1) present or (2) past owners of hazardous waste disposal facilities, (3) generators of hazardous waste, and (4) transporters of hazardous waste.
	Amended	1986		

The Emergency Planning and Community Right-to-Know Act (EPCRA)	1986	Title III of SARA contains a separate piece of legislation called EPCRA. There are two main goals of EPCRA: (1) to have states create local emergency units that must develop plans to respond to chemical release emergencies, and (2) to require the EPA to compile an inventory of toxic chemical releases to the air, water, and soil from manufacturing facilities.	Facilities must work with state and local entities to develop emergency response plans in case of an accidental release. Affected facilities must report to the EPA annually their data on the maximum amount of the toxic substance on-site in the previous year, the treatment and disposal methods used, and the amounts released to the environment or transferred off-site for treatment and/or disposal.
Pollution Prevention Act (PPA)	1990	The act established pollution prevention as the nation's primary pollution management strategy with emphasis on source reduction. It established a Pollution Prevention Information Clearinghouse whose goal is to compile source reduction information and make it available to the public.	The only mandatory provisions of the PPA require owners and operators of facilities that are required to file a Form R under the SARA Title III to report to the EPA information regarding the source reduction and recycling efforts that the facility has undertaken during the previous year.

along with a listing of some key requirements. A more complete description of these federal statutes is included in the appendix to this chapter.

3.3   EVOLUTION OF REGULATORY AND VOLUNTARY PROGRAMS FROM END-OF-PIPE TO POLLUTION PREVENTION AND SUSTAINABILITY

Many of the environmental laws that were reviewed in the previous section were enacted to ensure the protection of one specific environmental medium. For example, the Clean Air Act instituted a strategy for pollution control on sources that emit to the atmosphere. Similarly, the Clean Water Act and the Resource Conservation and Recovery Act provided systems for the protection of the water and land, respectively. Although these legislative actions have been extremely effective in cleaning up and controlling releases to certain environmental media, they did not initially ensure that the total amount of hazardous materials entering the environment would decrease. In fact, the volumes and hazards of toxic chemical releases into the environment continued to grow through the 1970s and 1980s, as the nation created and used more toxic chemicals (Johnson, 1992).

Beginning in the mid- to late 1980s, however, the absolute amounts of toxic releases to the environment, in many categories, began to decrease. If one uses the Toxics Release Inventory (TRI) as an indicator, the amount of "toxics" being released in the United States has decreased from 3.5 billion pounds per year to less than 2.0 billion pounds per year. The amount released has decreased every year from 1988 through the decade of the 1990s. These reductions have continued, although comparisons with previous years are made difficult as the scope of required reporting has changed. In 2009, TRI releases were roughly half of those reported in 1999. In addition, concentrations of many categories of pollutants in the environment are decreasing over time. This is true for ozone, lead, VOCs, and carbon monoxide (CO). Other environmental indicators are also showing improvement. For example, the amount of energy used per dollar of gross national product has decreased from about 10,800 BTU/2005 dollar in 1990 to 7300 in 2009 (EIA, 2011).

As traditional, media-specific regulations have reduced emissions to the environment, it has become more difficult to find additional opportunities for reductions in emissions and wastes. As additional reductions in emissions to individual environmental media are sought, it is important to guard against moving pollutants from one environmental medium into another. For example, traditional air pollution control devices, such as scrubbers, transfer pollutants from a gaseous stream to a liquid stream. The liquid stream would require further treatment to either remove or destroy the original contaminant. Similarly, some wastewater streams containing VOCs are contacted with an air stream, transferring the pollutants from the water to air. A more subtle form of media shifting can occur when pollutants are destroyed or transformed into less harmful forms by reaction during waste treatment. For example, disinfection of water through the use of chlorine can result in the creation and release of chlorinated organic compounds into the atmosphere, and thermal waste treatment processes can result in the formation and release of gases contributing to global warming when fuels are burned to combust the waste.

It is clear from the trends just discussed that a more effective strategy is needed to reduce the amounts and the hazardous characteristics of industrial wastes released into all media of the environment. This new strategy should also decrease the amounts of contaminants entering traditional waste treatment processes. In the next section of this chapter, we will review the environmental management hierarchy as outlined in the Pollution Prevention Act of 1990 and define important terms, such as pollution prevention, source reduction, and others.

3.4   POLLUTION PREVENTION CONCEPTS AND TERMINOLOGY

A logical starting point for understanding pollution prevention concepts is the waste management hierarchy established in the Pollution Prevention Act of 1990. The waste management hierarchy is defined as follows (42 U.S.C. §13101(b)):

> The Congress hereby declares it to be the national policy of the United States that pollution should be prevented or reduced at the source whenever feasible; pollution that cannot be prevented should be recycled in an environmentally safe manner, whenever feasible; pollution that cannot be prevented or recycled should be treated in an environmentally safe manner whenever feasible; and disposal or other release into the environment should be employed only as a last resort and should be conducted in an environmentally safe manner.

Based on this definition and distinctions between recycle options, we can place the waste management hierarchy in the following descending order, from the most to the least preferable:

1. Source reduction
2. In-process recycle
3. On-site recycle
4. Off-site recycle
5. Waste treatment
6. Secure disposal
7. Direct release to the environment

The distinction among these seven elements of the waste management hierarchy can be illustrated using examples from different engineering disciplines:

1. **Source reduction:** A metal machining operation reduces scrap waste by using computer-assisted machining to reduce operator error.
2. **In-process recycle:** In a chemical reactor, unreacted feed is separated and recycled back to the reactor.

3. **On-site recycle:** Waste heat from the ventilation exhaust of a building is transferred to incoming cold air, thus reducing energy use and emissions.

4. **Off-site recycle:** Coal fly ash from a power plant is transported to a cement manufacturer and mixed into new cement.

5. **Waste treatment:** Vapor emissions from tanks holding fuel at a petroleum refinery are captured and burned in a flare device.

6. **Secure disposal:** Sludge from a wastewater treatment plant is recovered and transferred to a landfill for disposal.

7. **Direct release to the environment:** Waste mine rock is separated from valuable ore and piled up on the surface of the Earth.

Other examples of pollution prevention through recycling can occur after product use, and engineers can have a great influence on the ability to recycle through design of the original product and also through design of recycling processes. For example, compact fluorescent lightbulbs (CFLBs) can be recycled after bulb use in order to recover and recycle mercury from the lamp and also to recycle metals, glass, and electronic components. The ability to recycle the CFLBs depends on the original bulb design and on the existence of engineered processes for bulb recycling. Additional examples of source reduction methods are available in several references (U.S. EPA, 1992, 1993; Hunt, 1995; Allen et al., 2002).

3.5   ENVIRONMENTAL LAW AND SUSTAINABILITY

Environmental laws are a starting point for legal frameworks addressing sustainability, and agencies such as the U.S. EPA are working to incorporate sustainability concepts into the way in which they implement environmental laws and regulations (NRC, 2011). However, laws are limited in extent by political boundaries, whereas environmental degradation is not. While much can be done within national borders, legal frameworks promoting sustainability will eventually need to be international in scope.

The United Nations provides one arena in which international agreements on sustainability can take shape. Much of the work of the UN is organized around Agenda 21, the Rio Declaration on Environment and Development. Agenda 21 is a plan of action for promoting sustainability, adopted by more than 170 governments at the United Nations Conference on Environment and Development (UNCED) held in Rio de Janeiro, Brazil, June 3–14, 1992. The UN Commission on Sustainable Development was created to ensure effective follow-up on UNCED.

The commission has developed programs, plans, and strategies (UN Commission on Sustainable Development, 2010a,b) and continues to hold regular sessions. For example, the 18th session, held in 2009 and 2010, focused on issues of transport, chemicals, waste management, and mining. The session also identified a ten-year framework of programs on sustainable consumption and production patterns and reviewed progress made in addressing the vulnerabilities of small island nations to rising sea levels (UN Commission on Sustainable Development, 2010a).

Progress has been slow on the development of binding international protocols since the Rio summit in 1992. However, the case of ozone-depleting substances, in which international protocols, which emerged rapidly after the discovery of polar ozone holes (see Chapter 1), limiting the manufacture and use of chlorofluorocarbons and other ozone-depleting substances is evidence that international agreements can be successful.

PROBLEMS

1. **Terms and definitions for environmental management** Provide definitions for the following terms:
   a. Pollution prevention
   b. Source reduction
   c. In-process versus on-site versus off-site recycling
   d. Waste treatment
   e. Disposal
   f. Direct release

2. **Solvent recovery operation in the automobile industry** Categorize the following solvent recovery operation in terms of the waste management hierarchy. Discuss the pollution prevention features of this process. Assess whether this process is pollution prevention.

   Process Description: The automotive industry uses robots to paint automobile bodies before attaching them to the chassis and installing other components such as

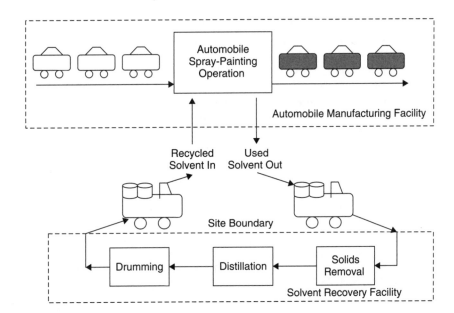

the drivetrain, lights, trim, and upholstery. In order to accommodate different colors, the paint lines must be flushed with a solvent and then recharged with the new color of paint. In the past, this solvent and paint residue was disposed of as hazardous waste or incinerated. The current process of spray painting automobiles uses a closed-loop solvent recovery process as outlined in the diagram on page 75 (Gage Products, Ferndale, MI).

3. **Analysis of federal environmental statutes** Choose one of the nine federal environmental statutes and analyze the regulatory provisions for the potential to impact a chemical production facility's capital and operating costs. What are the key provisions requiring action? What is the nature of those actions? What are the cost implications of those actions? The information contained in the appendix to this chapter will be helpful in answering these questions.

4. **U.S. Supreme Court decision on the Clean Air Act and greenhouse gases** There are eight sections to the U.S. Supreme Court ruling that the Clean Air Act provides the U.S. EPA with authority to regulate CO_2 and other greenhouse gases as air pollutants. Download a copy of this ruling and write short paragraph summaries of key concepts, developments, and rulings for each section. An electronic version may be found at www.law.cornell.edu/supct/pdf/05-1120P.ZO. This assignment may be suitable as a team project in which each student takes responsibility for one or more sections of the ruling.

5. **Superfund site investigation** Go to the EPA Superfund Web site (www.epa.gov/superfund/) and identify a Superfund site from those on the National Priority List. You may choose a site near your hometown or any on the list in which you have a special interest. Read the background on the site and summarize progress toward achieving the stated cleanup goals. Respond to this assignment using a one- to two-page memo format, single-spaced, including any figures or tables.

6. **Premanufacture Notice from the TSCA** Go to the EPA Web site (www.epa.gov/oppt/newchems/pubs/pmnforms.htm) and read about/summarize what information needs to be submitted about new chemical substances. Respond to this assignment using a one- to two-page memo format, single-spaced, including any figures or tables.

7. **Worker Protection Standard under FIFRA** Go to the EPA Web site for FIFRA (www.epa.gov/oecaagct/lfra.html) and review/summarize the Worker Protection Standard for Agricultural Pesticides. Respond to this assignment using a one- to two-page memo format, single-spaced, including any figures or tables.

8. **Green jobs and occupational safety issues** Go to the OSHA Web site (www.osha.gov/dep/greenjobs/index.html) and read about, then summarize, issues on green job hazards. Respond to this assignment using a one- to two-page memo format, single-spaced, including any figures or tables.

9. **Programs of the Clean Air Act** Go to the EPA Web site (www.epa.gov/air/caa/) and read about a current program or initiative in the Clean Air Act. One such program is the National Clean Diesel Initiative (www.epa.gov/diesel/). Summarize issues, programs, and outcomes. Respond to this assignment using a one- to two-page memo format, single-spaced, including any figures or tables.

10. **National Pollutant Discharge Elimination System of the Clean Water Act** The CWA established the National Pollutant Discharge Elimination System (NPDES) permit program that requires any point source of pollution to obtain a permit. Go to the EPA NPDES Web site (http://cfpub.epa.gov/npdes/) and read about and summarize

one of the Program Areas or Current Issues. Respond to this assignment using a one-
to two-page memo format, single-spaced, including any figures or tables.

11. **National Hazardous Waste Biennial Report of RCRA** The Resource Conservation
and Recovery Act (RCRA, accessible at the following Web site: www.epa.gov/
epawaste/inforesources/online/index.htm) publishes a *National Hazardous Waste
Biennial Report* (www.epa.gov/epawaste/inforesources/data/br09/index.htm) on the
state of hazardous waste generation, shipment, and management in the United States
on a state-by-state basis. Review the latest national analysis document. Summarize key
findings using a one- to two-page memo format, single-spaced, including any figures
or tables.

APPENDIX (ADAPTED FROM LYNCH, 1995)

The Toxic Substances Control Act (TSCA) of 1976

Incidents in which highly toxic substances, such as PCBs, began appearing in the environment and in food supplies prompted the federal government to create a program to
assess the risks of chemicals before they are introduced into commerce. The Toxic Substances Control Act (TSCA) was enacted on October 11, 1976. TSCA empowers the
EPA to screen new chemicals or certain existing chemicals to assure that their production and use does not pose "unreasonable risk" to human health and the environment.
However, TSCA requires the EPA to balance the economic and social benefits against
the purported risks.

Information Gathering

TSCA requires the EPA to gather information on all chemicals manufactured or processed for commercial purposes in the United States. The first version of the "TSCA Inventory" contained 55,000 chemicals. If a chemical is not found on this list, it is considered
to be a new chemical substance and is subject to the Premanufacture Notification requirements of Section 5 of TSCA. To aid in the gathering of information on existing
compounds, Section 8 of TSCA requires companies that manufacture, import, or process
any chemical substance to submit a report detailing chemical and processing information. This information includes the chemical identity, name, and molecular structure; categories of use; amounts manufactured or processed; by-products from manufacture,
processing, use, or disposal; environmental/health effects of chemical and by-products;
and exposure information. Companies must also keep records of any incidents involving
the chemical that resulted in adverse health effects or environmental damage.

Existing Chemicals Testing

TSCA may require companies to conduct chemical testing and then submit more detailed
data to the EPA. The EPA can request this additional data for chemicals that reside on a
separate list compiled from the TSCA Inventory by an Interagency Testing Committee.
Chemicals that become listed either are typically produced in very high volumes or may
pose unreasonable risk or injury to health or the environment. The list can contain no

more than 50 chemicals, and the EPA is required to recommend a test rule or remove the chemical from the list within one year of its listing. Once a test rule has been promulgated, a regulated entity has 90 days from the initiation of the test rule to submit the data.

New Chemical Review

Manufacturers, importers, and processors are required to notify the EPA within 90 days of introducing a new chemical into commerce by submitting a Premanufacture Notice (PMN). The PMN contains information on the identity of the chemical, categories of use, amounts intended to be manufactured, number of persons exposed to the chemical, the manner of disposal, and data on the chemical's effects on health and the environment. The EPA can require a PMN to be submitted on any existing chemical that is being used in a significantly different manner from prior known usages. The EPA has 90 days from the submission of the PMN to assess the risks of the new chemical or new usage of an existing chemical. If the risks are deemed to be unreasonable based on the information in the PMN and other data that are generally available, the EPA is required to take steps to control such risks. These steps might include limiting the production or use of the chemical or ruling a complete ban of the chemical. If data contained in the PMN are insufficient such that EPA cannot make a determination of the risks, the production of that chemical may be banned until such data are made available.

Regulatory Controls and Enforcement

The EPA has several options to control the risk of chemicals that have been deemed to pose unreasonable risk, ranging from banning the chemical (most burdensome), to limiting its production and use (less burdensome), to requiring warning labels at the point of sale (least burdensome). The EPA is required to use the least burdensome regulatory control considering the chemical's societal and economic benefits. This does not mean that the least burdensome control is always used, but rather it requires the EPA to consider the benefits before applying regulatory controls. The EPA is authorized to conduct inspections of facilities for manufacturing, processing, storing, or transporting regulated chemicals, and items eligible for inspection may include records, files, controls, and processing equipment. "Knowing or willful" violations of TSCA are punishable as crimes that carry up to one year imprisonment and up to $25,000 per day of violation.

The Federal Insecticide, Fungicide, and Rodenticide Act (FIFRA)

The Federal Insecticide, Fungicide, and Rodenticide Act was originally enacted in 1947 but has been amended several times, most notably in 1972 and 1996. Because all pesticides are toxic to plants and animals, they may pose an unacceptable risk to human health and the environment. FIFRA is a federal regulatory program whose purpose is to assess the risks of pesticides and to control their usage so that any exposure poses an acceptable level of risk.

Registration of Pesticides

Before any pesticide can be distributed or sold in the United States, it must be registered with the EPA. The decision by the EPA to register a pesticide is based on the data submitted by the pesticide manufacturer in the registration application. The data in the registration application are difficult and expensive to develop and must include the crop or insect to which the pesticide will be applied. In addition, the data must support the claim that the pesticide is effective against its intended target, that it allows adequate safety to those applying it, and that it will not cause unreasonable harm to the environment. The use of the term *unreasonable harm* is equivalent to requiring the EPA to consider a pesticide's environmental, economic, and social benefits and costs. Pesticides are registered for either general or restricted use. The EPA requires that restricted pesticides be applied by a certified applicator. A registration is valid for five years, at which time it automatically expires unless a reregistration petition is received. FIFRA requires older pesticides that were never subject to the current registration requirements to be registered if their use is to continue. It is estimated that there are over 35,000 older pesticides that were never registered during their prior usage. The EPA can cancel a pesticide's registration if the pesticide is found to present unreasonable risk to human health or the environment. Also, a registration may be revoked if the pesticide manufacturer does not pay the EPA the registration maintenance fee.

Labeling

Labels must be placed on pesticide products that indicate approved uses and restrictions. The label must contain the pesticide's active ingredients, instructions on approved applications to crops or insects, and any limitations on when and where it can be used. It is a violation of FIFRA to use any pesticide in a manner that is not consistent with the information contained on the product label.

Enforcement

It is unlawful to sell or distribute any unregistered pesticide or any pesticide whose composition or usage is different from the information contained in its registration. It is also a violation if FIFRA record-keeping, reporting, and inspection requirements are not met. The use of registered pesticides that were approved for restricted use only in any manner other than as stated on the FIFRA registration also constitutes a violation. Finally, it is unlawful to submit false data and registration claims. The power to enforce FIFRA is given to the states; however, the state implementation and enforcement programs must be substantially equivalent to the federal program. Any violation of FIFRA is punishable by a civil fine of up to $5000, and knowing violations of registration requirements may have criminal fines of up to $50,000 and one year imprisonment. Fraudulent data submissions may be punishable by up to $10,000 or up to three years imprisonment.

The Occupational Safety and Health Act (OSH Act) of 1970

The OSH Act was enacted on December 29, 1970, in order to assure safe working conditions for men and women. The agency that oversees the implementation of the OSH Act is the Occupational Safety and Health Administration (OSHA). Each state is authorized to develop its own safety and health plan, but it may adopt the federal program and must meet all federal standards. All private facilities having more than ten employees must comply with the OSH Act requirements, though certain employment sectors are exempt from the majority of the act's regulatory provisions. For example, excluded are certain segments of the transportation industry, which are covered by the Department of Transportation regulations; the mining industry, which is regulated by the Mine Safety and Health Administration; and the atomic energy industry, which must comply with the Nuclear Regulatory Commission standards.

Workplace Health and Safety Standards

The OSH Act requires OSHA to set workplace standards to assure a safe and healthy work environment. These include health standards, which provide protection from harmful or toxic substances by limiting the amount to which a worker is exposed, and safety standards, which are designed to protect workers from physical hazards, such as faulty or potentially dangerous equipment. When establishing health standards, OSHA considers the short-term (acute), long-term (chronic), and carcinogenic health effects of a chemical or a chemical mixture. These standards take the form of maximum exposure concentrations for chemicals and requirements for labeling, use of protective equipment, and workplace monitoring.

Hazard Communication Standard

The OSH Act's Hazard Communication Standard requires that several standards be met by manufacturers or importers of chemicals and also for the subsequent users of them. These requirements include the development of hazard assessment data, the labeling of chemical substances, and the informing and training of employees in the safe use of chemicals. Manufacturers and importers are required to assess both the physical and health hazards of the chemicals they make or use. This information must be assembled in a material safety data sheet (MSDS) in accordance with OSH Act standards and accompany any sale or transfer of the chemical. Chemical manufacturers and importers must also label chemicals according to OSH Act standards whenever a chemical leaves their control and must train their employees on the safe handling of chemicals in the workplace. Employers must keep a copy of the MSDS in the workplace for each chemical used. Employers must also develop a written hazard communication plan that outlines the implementation plan for informing and training employees on the safe handling of chemicals in the workplace. Employers that use manufactured chemicals must also label those containers according to OSH Act standards.

Record-Keeping and Inspection Requirements

Employers must keep records of all steps taken to comply with OSH Act requirements, including the company's safety policies, hazard communication plan, and employee training programs. In addition, employers must keep records of all work-related injuries and deaths and report them periodically to OSHA. Employers must keep records of employee exposure to potentially toxic chemicals for 30 years. An OSHA Compliance Safety and Health Officer is authorized to enter all covered facilities as part of a general inspection schedule in order to review safety policies and records and to inspect manufacturing equipment. After inspection, a closing meeting is held between the inspector and company health and safety representatives to discuss any potential OSH Act violations.

Enforcement

Based on the inspection, a citation may be issued for any OSH Act violations. These citations must be posted in a prominent location within facilities for at least three days. *De minimis* violations are not considered serious enough to threaten employee safety and health. Serious violations present a real potential for employee harm and may involve penalties of up to $7000. Willful or repeated violations carry penalties of up to $70,000 per violation.

Clean Air Act (CAA) of 1970

The Clean Air Act is actually an amendment of an earlier law (the 1955 Air Pollution Act had weak regulatory provisions) and has been amended multiple times, most notably in 1977 and 1990. The CAA is intended to control the discharge of air pollution by establishing uniform ambient air quality standards that are in some instances health-based and in others technology-based. Mobile and stationary sources of air pollution must comply with source-specific emission limits that are intended to meet these ambient air quality standards. In addition, the CAA addresses specific air pollutants and air pollution problems such as hazardous air pollutants, stratospheric ozone depletion, and acid rain. The Supreme Court has ruled that the act also allows for regulation of greenhouse gases. The 1990 amendments of the CAA revised the hazardous air pollutant regulatory program, instituted a market-based emissions trading system for sulfur dioxide, created strict tailpipe emission standards for the most polluted urban areas, created a market for reformulated and alternative fuels, and instituted a comprehensive state-run operating permit program.

One of the most important steps in achieving the goals of the CAA was the establishment of the National Ambient Air Quality Standards (NAAQS). These are the maximum allowable concentrations of specific chemicals monitored in the ambient, or background, air that meet or exceed health-based criteria. Primary and secondary NAAQS are set for the criteria pollutants, carbon monoxide, lead, nitrogen dioxide, tropospheric ozone, particulate matter, and sulfur dioxide. The NAAQS primary standards are human-health-related, and the secondary standards are intended to prevent a broader range of environmental harm (soils, crops, vegetation, and wildlife); thus they are more restrictive than the primary standards.

State Implementation Plan

The CAA requires states to develop individualized state implementation plans (SIPs) that outline how they intend to achieve the NAAQS. The SIP-NAAQS system is an example of "cooperative federalism." The federal government assures that the provisions of the CAA are implemented, but states are responsible for controlling local sources of air pollution. Thus, the state regulatory agencies establish source-specific emission limits on mobile and stationary sources at a sufficient level to ensure compliance with federal quality standards. Under the CAA, the EPA establishes the NAAQS, reviews state-authored SIPs to ensure that they will achieve the NAAQS, and may take over state programs if they fail to implement the SIP effectively.

New Source Performance Standards

The CAA allows emission limits to be set on new sources that are more restrictive than limits on existing sources. These standards are termed *new source performance standards* (NSPS). The reasoning behind these standards is that it is easier to incorporate controls into new processes than to retrofit them into existing processes. The EPA established which categories of industrial sources can be subject to these standards, and the emission limits are set by considering the best available emission control technologies, other health and environmental impacts that may occur during the application of the control technology, and energy usage issues. Because the new source standards are uniformly established nationwide, they create a level playing field where companies are discouraged from locating in states that do not require these strict pollution controls.

A New Source Review Program has been established by the CAA in order to review new processes and significant modifications to existing processes and to prevent significant deterioration of ambient air quality. Before construction can begin, the operator must obtain a permit and demonstrate that (1) the source will comply with ambient air quality standards, (2) the source will use the best available control technology, (3) the emissions will not cause a violation of the NAAQS in nearby areas, and (4) new or modified sources will achieve offsets, that is, reductions in emissions of the same pollutant, in a greater than one-to-one ratio.

Hazardous Air Pollutants

The CAA has identified more than 180 hazardous air pollutants (HAPs) that are subject to more stringent emission controls than the six criteria air pollutants. Any stationary source that emits 10 tons per year of any HAP or 25 tons per year of any combination of HAPs is subject to these CAA provisions. The EPA is required to develop source-specific emission standards that require installation of technologies that will result in the maximum achievable degree of control (MACT). If an existing source can demonstrate that it has achieved or will achieve a reduction of 99% of hazardous air pollution emissions before enactment of the MACT standards, it may receive a six-year extension of its compliance deadline.

Greenhouse Gases

In the near future, the U.S. EPA will decide whether and how to regulate greenhouse gases as air pollutants under the CAA. As of this writing, three major developments have occurred (Solomon, 2009). First, in the Supreme Court case of *Massachusetts v. EPA*, the court ruled in a 5-to-4 decision that the EPA has the statutory authority to regulate CO_2 and other greenhouse gases as air pollutants under the Clean Air Act (Cornell University, 2007). Second, on September 22, 2009, the EPA administrator signed the *Final Mandatory Reporting of Greenhouse Gases Rule*. Under this rule large emission sources and suppliers are required to report greenhouse gas emissions each year for facilities that emit greater than 25,000 metric tons annually (U.S. EPA, 2009a). The intention of the rule is to collect accurate data for future policy decision making on climate change mitigation. Third, in December 2009 the EPA made a finding that greenhouse gases endanger human health and welfare, in response to the 2007 Supreme Court ruling (U.S. EPA, 2009b). These gases include carbon dioxide (CO_2), methane (CH_4), nitrous oxide (N_2O), hydrofluorocarbons (HFCs), perfluorocarbons (PFCs), and sulfur hexafluoride (SF_6). This endangerment finding is a prerequisite to the EPA developing emission standards for greenhouse gases.

Enforcement

Civil penalties for violations of the Clean Air Act may involve fines of up to $25,000 per day of violation. Fines for knowing violations of the CAA are up to $250,000 per day and up to five years imprisonment. Corporations may be fined up to $500,000 per violation and repeat offenders may receive double fines. Knowing violations that involve releases of HAPs may trigger fines of up to $250,000 per day and up to 15 years imprisonment. Corporations may be fined $1,000,000.

The Clean Water Act (CWA) of 1972

The Clean Water Act (CWA) was first enacted on October 18, 1972, and is the first comprehensive federal program designed to reduce pollutant discharges into the nation's waterways ("zero discharge" goal). Another goal of the CWA is to make water bodies safe for swimming, fishing, and other forms of recreation ("swimmable" goal). This act has resulted in significant improvements in the quality of the nation's waterways since its enactment. The CWA defines a pollutant rather broadly, as "dredged spoil, solid waste, incinerator residue, sewage, garbage, sewage sludge, munitions, chemical wastes, biological materials, radioactive materials, heat, wrecked or discarded equipment, rock, sand, cellar dirt and industrial, municipal, and agricultural waste discharged into water" (CWA §502(14), 33 U.S.C. §1362). The CWA has two major components, the National Pollutant Discharge Elimination System (NPDES) permit program and the Publicly Owned Treatment Works (POTW) construction program.

Publicly Owned Treatment Works (POTW) Construction Program

This program originally provided grants to POTWs so that they could upgrade their facilities from primary to secondary treatment. Primary treatment involves removing a portion of the suspended solids and organic matter using operations such as screening and sedimentation. Secondary treatment removes residual organic matter using microorganisms in large mixed basins. Federal grants having no repayment obligations were available for as much as 55% of the total project costs. The 1987 amendments converted the grant program into a revolving loan program in which municipalities can obtain low-interest loans that must be repaid.

National Pollutant Discharge Elimination System (NPDES) Permit Program

The statute classifies water pollution sources as point sources and non-point sources. Point sources are any discrete conveyances (pipes or ditches) that introduce pollutants into a water body. Point sources are further divided into municipal (from POTWs) and industrial. An example of a non-point source is runoff from agricultural lands. Non-point sources are the last major source of uncontrolled pollution discharge into waterways. The NPDES permit program requires any point source of pollution to obtain a permit. It is another example of a cooperative federal-state regulatory program. The federal government established national standards (e.g., effluent guidelines), and the states are given flexibility in achieving these standards. NPDES permits contain effluent limits, requiring either the installation of specific pollutant treatment technologies or adherence to specified numerical discharge limits. In establishing the NPDES limits, the state regulatory agency considers the federal effluent guidelines and the desired water quality standards established by the state for the intended use of the waterway (drinking water source, recreation, agricultural, etc.).

Monitoring/Inspection Requirements

NPDES permit holders must monitor discharges, collect data, and keep records of the pollutant levels of their effluents. These records must be submitted to the agency that granted the NPDES permit in order to assure that the point source is not exceeding the effluent discharge limits. The permitting agency is authorized to inspect the permit holders' records and collect effluent samples to verify compliance with the CWA.

Industrial Pretreatment Standards

Industrial sources that discharge into sewers that eventually enter POTWs are termed *indirect discharge* sources. These sources do not need to obtain an NPDES permit but may have to apply for state or local permits and must comply with EPA pretreatment standards. Pretreatment standards that reflect the best available control technology (BACT) are designed to remove the most toxic pollutants and to minimize the "pass-through" of these components into receiving waters from POTWs. Indirect dischargers can obtain removal credits if they can demonstrate that the POTW can effectively reduce a particular pollutant to acceptable levels.

Dredge and Fill Permits and Discharge of Oil or Hazardous Substances

A permit must be obtained from the United States Army Corp of Engineers before any discharge of dredge or fill materials occurs into navigable waterways, including wetlands. The CWA also prohibits discharge of oil or hazardous substances into any navigable waters and provides mechanisms for the cleanup of oil and hazardous substance spills. Any person in charge of a vessel or facility must notify the Coast Guard's National Response Center and also state officials whenever such a spill occurs above a certain quantity. Failure to do so may result in up to five years imprisonment.

Enforcement

Civil penalties may be as high as $25,000 per day for violations of the CWA provisions. Criminal violations for repeated negligent conduct may be as high as $50,000 per day and up to two years imprisonment. Repeated knowing violations can result in fines of up to $100,000 per day and six years imprisonment. Repeated knowing endangerment violations of the CWA can bring fines as high as $500,000 and 15 years imprisonment. Organizations can be fined as much as $1,000,000. Violations that involve false monitoring and reporting are subject to a $10,000 fine and up to two years imprisonment.

Resource Conservation and Recovery Act (RCRA) of 1976

The Resource Conservation and Recovery Act was enacted to regulate the disposal of both non-hazardous and hazardous solid wastes to land, encourage recycling, and promote the development of alternative energy sources based on solid waste materials. In reality, RCRA also regulates any waste material that is disposed to land, including liquids, gases, and mixtures of liquids with solids and gases with solids. RCRA's Subtitle C provisions regarding the management and disposal of hazardous wastes have become the key provisions. RCRA was significantly amended by the Hazardous and Solid Waste Amendments (HSWA) in 1984. The provisions of the HSWA affect hazardous waste disposal facilities by restricting the disposal of hazardous waste, and they regulate underground storage tanks containing hazardous substances or petroleum. RCRA's Subtitle C establishes provisions that must be complied with by hazardous waste generators, transporters of hazardous waste, and facilities that treat, store, or dispose of hazardous waste. RCRA represents a "cradle-to-grave" regulatory system that manages hazardous waste throughout its life cycle in order to minimize the risks that these wastes pose to the environment and to human health.

Identification/Listing of Hazardous Waste

If wastes exhibit any of four hazardous characteristics (ignitability, corrosivity, reactivity, or toxicity), they are considered to be hazardous. A material can also be designated as a hazardous waste if the EPA lists it as such. Three hazardous waste lists have been compiled by the EPA. The first list contains approximately 500 wastes from nonspecific sources and includes specific chemicals. The second list of hazardous wastes is from specific industry sources, for example, hazardous wastes from the petroleum-refining industry. The third list includes wastes from commercial chemical products, which when discarded or spilled must be managed as hazardous wastes. Specifically exempted from being hazardous wastes are household waste, agricultural wastes that are returned to the ground as fertilizer, and wastes from the extraction, beneficiation, and processing of ores and minerals, including coal.

Generator Requirements

The EPA defines a generator as any facility that causes the generation of a waste that is listed as a hazardous waste under RCRA provisions. A generator of hazardous waste must obtain an EPA identification number within 90 days of the initial generation of the waste. RCRA requires generators to properly package hazardous waste for shipment off-site and to use approved labeling and shipping containers. Generators must maintain records of the quantity of hazardous waste generated as well as where the waste was sent for treatment, storage, or disposal and file these data in biennial reports to the EPA. Generators must prepare a Uniform Hazardous Waste Manifest, which is a shipping document that must accompany the waste at all times. A copy of the manifest will be sent back to the generator by the treatment facility to assure that the waste reached its proper destination.

Other Requirements

RCRA imposes requirements on transporters of hazardous waste as well as on facilities that treat, store, and dispose of hazardous wastes. Transporters are any persons who transport hazardous waste by air, rail, highway, or water from the point of generation to the final destination of treatment, storage, or disposal. The final destinations are termed *treatment, storage, and disposal facilities* (TSDFs) by the EPA. Transporters must adhere to the Uniform Hazardous Waste Manifest system when shipping hazardous waste, which includes retaining copies of manifests for a period of three years. A facility that accepts hazardous waste for the purpose of changing the physical, chemical, or biological character of the waste and with the intent of rendering the waste non-hazardous, making the waste amenable for transport or recovery, or reducing the waste volume is defined as a treatment facility by RCRA. Storage facilities are intended for holding wastes for a short period of time until such time as the waste is shipped to a treatment or disposal facility elsewhere. A disposal facility is a location that is engineered to safely accept hazardous waste in various forms (drums, solids, etc.) for long-term internment. These facilities must monitor the environment within and adjacent to the facility to assure that hazardous waste components are not leaving the site in concentrations that threaten the environment or human health. Generators who store hazardous waste on-site for more than 90 days or who treat or dispose of hazardous waste themselves are considered TSDFs by RCRA.

Enforcement

Failure to comply with RCRA Subtitle C or EPA compliance orders carries a civil penalty of up to $25,000 per day of violation. Violations may result in the revocation of the RCRA permit. Criminal penalties for violations may be as high as $50,000 per day for each violation and/or two years imprisonment. Fines and jail time may double for repeat offenders. When a person violates RCRA and in the process knowingly endangers another individual, fines may reach $250,000 per day and up to 15 years imprisonment. Organizations may be fined as much as $1,000,000.

The Comprehensive Environmental Response, Compensation, and Liability Act (CERCLA) of 1980

The contamination of Love Canal in upstate New York with industrial toxic materials and the subsequent evacuation of hundreds of families from the vicinity alerted the federal government to the need to clean up this and other related sites. The Comprehensive Environmental Response, Compensation, and Liability Act (CERCLA) of 1980 began a process of identifying and cleaning up the many sites of uncontrolled hazardous waste disposal at abandoned sites, at industrial complexes, and at federal facilities. The EPA is responsible for creating a list of the most hazardous sites of contamination, which is termed the National Priority List (NPL). As of 2011, there were 1,112 facilities, including 158 federal facilities, on the NPL, and an additional 62 sites are proposed for addition to the NPL. CERCLA established a $1.6 billion Hazardous Substance Trust Fund (Superfund) to initiate cleanup of the most contaminated sites. Superfund (the trust fund) allows for the cleanup of sites for which parties responsible for creating the contamination cannot be identified because of bad record keeping in the past, or are no longer able to pay, are bankrupt, or are no longer in business. The Superfund Amendments and Reauthorization Act (SARA) of 1986 increased the Superfund appropriation to $8.5 billion through December 31, 1991, extended and expanded the tax for Superfund, and stipulated a preference for remedial action to be cleanup rather than containment of hazardous waste. In addition, Superfund was extended to September 30, 1994, with an additional $5.1 billion. As of this writing, the Superfund trust fund has been exhausted and the Superfund program continues to operate via yearly U.S. EPA budget appropriations, fund interest, and cost recoveries from PRPs (see the next section), though no new appropriations have been added to the trust fund since 1995. Under the CERCLA provisions, the EPA can respond to sites of hazardous waste contamination in two ways: (1) short-term emergency responses to spills or other releases, and (2) long-term remedial actions, which may actually occur long after the site is listed on the NPL, and which are designed to achieve a permanent state of cleanup.

Potentially Responsible Party (PRP) Liability

After a site is listed in the NPL, the EPA identifies potentially responsible parties (PRPs) and notifies them of their potential CERCLA liability. If the cleanup is conducted by the EPA, the PRPs are responsible for paying their share of the cleanup costs. If the cleanup has not begun, PRPs can be ordered to complete the cleanup of the site. PRPs are (1) present or (2) past owners of hazardous waste disposal facilities, (3) generators of hazardous waste who arrange for treatment or disposal at any facility, and (4) transporters of hazardous waste to any disposal facility. Liability for PRPs is strict, meaning that liability can be imposed regardless of fault or negligence. Liability is joint and several, meaning that one party can be held responsible for the actions of others when the harm is indivisible. Finally, the liability is retroactive, meaning that parties can be held liable for actions that predate CERCLA. The EPA does not have to prove that a particular PRP's waste caused the contamination. EPA only has to prove that there are hazardous substances present at the site that are similar to those associated with a party's hazardous waste treatment and disposal activities.

Enforcement

The EPA can force PRPs to conduct and fund cleanup of contaminated sites with which they have been associated in actions termed Private Party Cleanups. Failure to comply with a Private Party Cleanup order may involve fines of up to $25,000 per day, and judicial review of these cases is not immediately available. Thus, PRPs have little choice but to comply. Failure to report to the EPA the release of a hazardous substance in quantities greater than the cutoff value for that substance may result in a fine amounting to more than $25,000 per day and criminal penalties of three years for a first conviction and five years for a subsequent conviction.

The Emergency Planning and Community Right-to-Know Act (EPCRA)

In 1984, the release of methyl isocyanate from a Union Carbide plant in Bhopal, India, killed more than 2500 people and permanently disabled some 50,000 more. This unfortunate incident illustrated the need for communities to develop emergency plans in preparation for releases that might occur from chemical manufacturing facilities. It also highlighted the need for communities to find out what toxic chemicals are being manufactured at facilities and to what media toxic chemicals are being released. Title III of SARA contains a separate piece of legislation called the Emergency Planning and Community Right-to-Know Act (EPCRA). There are two main goals of EPCRA: (1) to have states create local emergency units that must develop plans to respond to chemical release emergencies, and (2) to require the EPA to compile an inventory of toxic chemical releases to the air, water, and soil from manufacturing facilities, and to disclose this inventory to the public.

Toxics Release Inventory (TRI)

EPCRA requires facilities with more than ten employees that either use more than 10,000 pounds or manufacture or process more than 25,000 pounds of one of the listed chemicals or categories of chemicals to report annually to the EPA. The report must contain data on the maximum amount of the toxic substance on-site in the previous year, the treatment and disposal methods used, and the amounts released to the environment or transferred off-site for treatment and/or disposal. Facilities that are obligated to report must use the Chemical Release Inventory Reporting Form (Form R). Facilities must keep records supporting their TRI submissions for three years from the date of submission of Form R to the EPA. The data are compiled by the EPA and entered into a computerized database that is accessible to the public. The TRI is viewed by citizens, environmental groups, states, industry, and others as an environmental scorecard for industrial facilities.

Enforcement

Violations of EPCRA's TRI reporting and community emergency planning requirements are subject to civil penalties of up to $25,000 per day. Any person who knowingly and willingly fails to report releases of toxic substances can be fined up to $25,000 and/or be imprisoned for up to two years. Second violations may subject persons to fines of up to $50,000 or five years imprisonment.

Pollution Prevention Act of 1990

On October 27, 1990, Congress passed the Pollution Prevention Act (PPA), which established pollution prevention as the nation's primary pollution management strategy. Pollution prevention is defined as "any practice which: 1) reduces the amount of any hazardous substance, pollutant, or contaminant entering any waste stream or otherwise released into the environment prior to recycling, treatment, and disposal: and 2) reduces the hazards to public health and the environment associated with the release of such substances, pollutants, or contaminants." Thus, pollution prevention not only encourages reductions in waste generation and release from production facilities but also promotes reductions in waste component toxicity or other hazardous characteristics. This strategy is fundamentally different from that of prior environmental statutes, in that pollution prevention encourages steps to reduce pollution generation and toxicity at the source rather than relying on end-of-pipe pollution controls.

The PPA provides for a hierarchy of pollution management approaches. It states that (1) pollution should be prevented or reduced at the source whenever feasible, (2) pollution that cannot be prevented or reduced should be recycled, (3) pollution that cannot be prevented or reduced or recycled should be treated, and (4) disposal or other releases into the environment should be employed only as a last resort. The act is not an action-forcing statute but rather encourages voluntary compliance by industry of the suggested approaches and strategies through education and training. To this end, the EPA is required to establish a Pollution Prevention Office independent of the other media-specific pollution control programs. It is also required to set up a Pollution Prevention Information Clearinghouse whose goal is to compile source reduction information and make it available to the public. The only mandatory provisions of the PPA require owners and operators of facilities that are required to file a form R under SARA Title III (the TRI) to report to the EPA information regarding the source reduction and recycling efforts that the facility has undertaken during the previous year.

REFERENCES

Allen, D. T., and D. R. Shonnard. 2002. *Green Engineering: Environmentally Conscious Design of Chemical Processes*. Upper Saddle River, NJ: Prentice Hall.

Allen, D. T., D. Bauer, B. Bras, T. Gutowski, C. Murphy, T. Piwonka, P. Sheng, J. Sutherland, D. Thurston, and E. Wolff. 2002. "Environmentally Benign Manufacturing: Trends in Europe, Japan, and the USA." *ASME Journal of Manufacturing Science and Engineering* 124:908–20.

Cornell University, Law School Legal Information Institute. 2007. Supreme Court of the United States, *Massachusetts et al., Petitioners v. Environmental Protection Agency et al.* www.law.cornell.edu/supct/pdf/05-1120P.ZO. Accessed December 29, 2009.

EIA (Energy Information Administration, U.S. Department of Energy). 2011. *International Energy Outlook—2011*. www.eia.gov/countries/. Accessed March 2011.

Hunt, G. E. 1995. "Overview of Waste Reduction Techniques Leading to Pollution Prevention." In *Industrial Pollution Prevention Handbook*, edited by H. M. Freeman. New York: McGraw-Hill, pp. 9–26.

Johnson, S. 1992. "From Reaction to Proaction: The 1990 Pollution Prevention Act." *Columbia Journal of Environmental Law* 17:153.

Lynch, H. 1995. *A Chemical Engineer's Guide to Environmental Law and Regulation*. Ann Arbor, MI: National Pollution Prevention Center for Higher Education, University of Michigan.

NRC (National Research Council). 2011. *Sustainability and the U.S. EPA*. Washington, DC: National Academy Press.

Solomon, B. D. 2009. Personal communication, Michigan Technological University, December 29.

Sullivan, T. F. P., and T. L. Adams. 1997. *The Environmental Law Handbook*. Rockville, MD: Government Institutes.

UN (United Nations) Commission on Sustainable Development. 2010a. *Report on the Eighteenth Session*. E/CN.17/2010/15 E/2010/29. Available at www.un.org/esa/dsd/resources/res_docucsd_18.shtml. Accessed July 2011.

————. 2010b. *Milestone Reports*. Available at www.un.org/esa/dsd/dsd/dsd_milestones.shtml. Accessed July 2011.

U.S. EPA (U.S. Environmental Protection Agency), Office of Research and Development. 1992. *Pollution Prevention Case Studies Compendium*. EPA/600/R-92/046. April.

————. 1993. *DuPont Chambers Works Waste Minimization Project*. EPA/600/R-93/203. November, pp. 86–91.

————. 2009a. *Final Mandatory Reporting of Greenhouse Gases Rule*. www.epa.gov/climatechange/emissions/ghgrulemaking.html. Accessed December 29, 2009.

————. 2009b. *Endangerment and Cause or Contribute Findings for Greenhouse Gases under the Clean Air Act*. www.epa.gov/climatechange/endangerment.html. Accessed December 29, 2009.

West Law School. 2011. *Selected Environmental Law Statutes, 2011–2012, Educational Edition*. West Law Publishing.

CHAPTER 4

Green, Sustainable Materials

4.1 INTRODUCTION

Almost every engineering design involves the use of materials. If an engineering design is to be as sustainable as possible, the materials that are involved in the embodiment of the design should have light environmental and natural resource use footprints. Determining the magnitude of a material's footprint is not straightforward, however. In this chapter, a three-pronged approach to characterizing the footprint of materials will be described.

To be used in an engineering design, a material must first be extracted and purified (refined), and the overall impact of this extraction and refining is the first component of the footprint. Once the material enters the production system, it can be reused and recycled, reducing the need for extraction processes. The extent to which various materials are reused and recycled, and the energy and other resources used in processing the materials, is a second component of the footprint. Finally, if a material is not reused or recycled but escapes into the environment, its environmental fate, persistence, human health impact, and ecological impact are a third component of the footprint. Methods for characterizing the environmental footprints associated with each of these steps (extraction, processing, and environmental release) are described in the next three sections. The chapter concludes by discussing methods for combining these assessments.

4.2 ENVIRONMENTAL AND NATURAL RESOURCE USE FOOTPRINTS OF MATERIAL EXTRACTION AND REFINING

One of the simplest approaches for characterizing the footprints associated with the extraction and production of a material is to assess the material's overall scarcity. Simply stated, if a material is scarce, it is likely to be energy-intensive to obtain and refine it, and the ability to meet large-scale demand will be limited. Elements vary widely in their natural abundance. Table 4-1 reports the relative abundance of materials in the Earth's crust (note that this does not include oceans). The most common elements on a mass basis are, in descending order, O, Si, Al, Fe, Ca, Mg, and K. All of these elements are present at greater than 1% abundance in the Earth's crust, and all are present in widely used commodity materials. In contrast, some widely used elements (Ag, Sn, Sb) are present at the ppm level or lower, on average, in the crust. Figure 4-1 shows the relative amounts of various elements that are produced each year for commercial applications (Gerst and Graedel, 2008).

The extent to which elements are extracted can be quantitatively compared to the crustal abundance to produce one measure of the supply of the material. Example 4-1 examines the ratio of total crustal abundance to annual production rate. The result of the calculation in Example 4-1, the number of years at which current consumption rates could be sustained if all of the material in the Earth's crust could be extracted, is a very simplistic measure of material supply. Not all of the elemental material in the Earth's crust can be effectively extracted. Much of the material is simply present at concentrations that are too low to be extracted cost-effectively. Figure 4-2 shows the

Table 4-1   Abundance of Selected Elements in the Earth's Crust (Mass Abundance)

Element	Abundance	Element	Abundance	Element	Abundance
O	46.4%	Rb	90 ppm	U	2.7 ppm
Si	28.2%	Ni	75 ppm	Sn	2 ppm
Al	8.2%	Zn	70 ppm	Ta	2 ppm
Fe	5.6%	Ce	60 ppm	As	1.8 ppm
Ca	4.1%	Cu	55 ppm	Mo	1.5 ppm
Na	2.4%	Y	33 ppm	W	1.5 ppm
Mg	2.3%	Nd	28 ppm	Sb	0.2 ppm
K	2.1%	Co	25 ppm	Cd	0.2 ppm
Ti	0.6%	Sc	22 ppm	Bi	0.17 ppm
P	0.1%	Li	20 ppm	Pd	0.15 ppm
Mn	0.1%	N	20 ppm	In	0.1 ppm
Fl	0.06%	Nb	20 ppm	Hg	0.08 ppm
Ba	0.04%	Ga	15 ppm	Ag	0.07 ppm
Sr	0.04%	Pb	12 ppm	Se	0.005 ppm
S	0.03%	B	10 ppm	Pt	0.005 ppm
C	0.02%	Th	10 ppm	Au	0.004 ppm
Zr	0.02%				
V	0.01%				
Cl	0.01%				
Cr	0.01%				

Source: Taylor, 1964

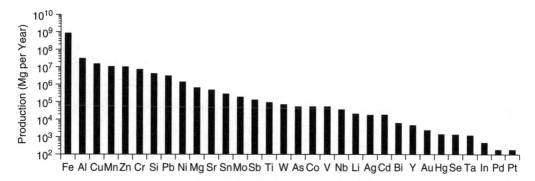

Figure 4-1 Annual production of elements for use in commerce (Reprinted with permission from Gerst and Graedel, 2008. Copyright 2008 American Chemical Society)

relationship between total crustal abundance and identified and economically extractable deposits (reserves). On average, only 1 in 10^7 to 10^9 tons of an element in the Earth's crust is an economically viable reserve of the material. Example 4-2 examines the relationship between reserves and annual use rates for several commodity materials.

Example 4-1   Supplies and use of elements

Calculate the ratio of the abundance of materials in the Earth's crust (tons) to the annual use of the materials (tons/yr). Assume an approximate mass for the crust of

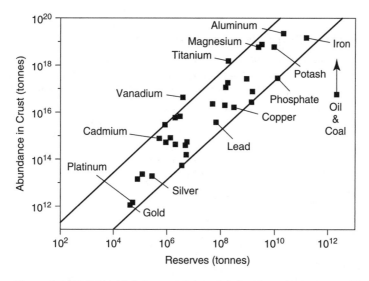

Figure 4-2 Relationship between total crustal abundance and reserves (Kesler, 1994)

$2 * 10^{19}$ metric tons. This mass is based on a 40 km thickness for continental crust and a 3 km thickness for oceanic crust, with an average density of 3 g/cm³ (3 tons/m³).

Element	Abundance in Crust	Use (tons/yr)	Ratio of Abundance to Use (yr)
Fe	5.6%	1,200,000,000	
P	0.1%	153,000,000	
Ni	75 ppm	1,300,000	
Zn	70 ppm	12,500,000	
Cu	55 ppm	15,000,000	
Ag	0.07 ppm	23,000	

Solution:

For example, for silver:

Abundance in crust * crustal mass/use rate = $7 * 10^{-8} * 2 * 10^{19}$ metric tons/ $(2.3 * 10^4) = 6 * 10^7$ yr

Element	Abundance in Crust	Use (tons/yr)	Ratio of Abundance to Use (yr)
Fe	5.6%	1,200,000,000	$1*10^9$
P	0.1%	153,000,000	$1*10^8$
Ni	75 ppm	1,300,000	$1*10^9$
Zn	70 ppm	12,500,000	$1*10^8$
Cu	55 ppm	15,000,000	$7*10^7$
Ag	0.07 ppm	23,000	$6*10^7$

Example 4-2   Economically extractable supplies and use of materials

Calculate the ratio of the economically extractable materials in the Earth's crust (reserves) to the annual use of the materials (tons/yr). The quantities of reserves identified in this exercise are derived from the U.S. Geological Survey (USGS, 2010).

Element	Economically Extractable Resources (tons)	Use (tons/yr)	Ratio of Abundance to Use (yr)
Fe	230,000,000,000	1,200,000,000	
Ni	130,000,000	1,300,000	
Zn	1,900,000,000	12,500,000	
Cu	3,000,000,000	15,000,000	
Ag	400,000	23,000	
Sb	2,100,000	200,000	
Sn	5,600,000	300,000	

Solution:

Element	Economically Extractable Resources (tons)	Use (tons/yr)	Ratio of Abundance to Use (yr)
Fe	230,000,000,000	1,200,000,000	190
Ni	130,000,000	1,300,000	100
Zn	1,900,000,000	12,500,000	152
Cu	3,000,000,000	15,000,000	200
Ag	400,000	23,000	17
Sb	2,100,000	200,000	10
Sn	5,600,000	300,000	19

As illustrated in Example 4-1, the total amount of material available in the Earth's crust is sufficient to support current rates of extraction indefinitely. However, Example 4-2 shows that the total amount of crustal material that is present in high enough concentrations to be economically recoverable is much lower than the total amount of material. This is because energy and other resources are required to purify materials into commercially useful forms. An approximate relationship between the concentration at which a material is found in a raw material and the cost to refine the material is shown in Figure 4-3. This concept—that the cost of a material is largely determined by the cost of extracting and purifying the raw

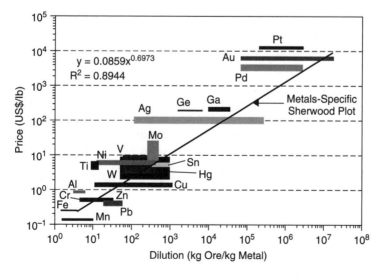

Figure 4-3 Metal prices (2004) as a function of dilution (1/concentration) of metals in commercial ores; the relationship illustrates the concept that the more dilute a material is in its native ore, the more expensive it will be to purify into a commodity material. (Reprinted with permission from Johnson et al., 2007. Copyright 2007 American Chemical Society)

material—is sometimes referred to as the *Sherwood relationship*, named after chemical engineering Professor T. Sherwood of MIT, who noted the relationship in the 1950s. Johnson et al. (2007) updated the data for the 21st century to create the information shown in Figure 4-3.

The relationship between prices and dilution in ores—the Sherwood relationship—is technology-dependent. As technologies continue to become more efficient, costs associated with material extraction can be reduced. There is, of course, a physical limit. A certain amount of entropy must be overcome to purify a material, and that imposes a minimum energy burden that must be overcome. As Example 4-3 shows, however, current technologies are far from those entropy limits.

Example 4-3   Energy burdens of material recovery (adapted from Allen, 1996)

To recover resources from raw materials, the minimum amount of energy that must be invested to concentrate the material is determined by the entropy that must be overcome in the purification.

a. To make an estimate of this energy, calculate the entropy of mixing that must be overcome in concentrating 1 kg of a metal present at 0.2 ppm (mole fraction basis) in water (imagine a process that seeks to harvest lithium from seawater to manufacture lithium-ion batteries). The entropy of mixing is given by

$$\Delta S = -R \, \Sigma \, x_i \ln x_i$$

where ΔS is the molar entropy of mixing, R is the gas constant, x_i is the mole fraction of each component in the mixture, and the summation is done over all components in the mixture. Lithium has an atomic weight of 6.94.

b. The minimum energy required for separation (the energy required to overcome changes in entropy) is given by

$$\Delta G = T\Delta S$$

where ΔG is the Gibbs free energy of mixing and T is the absolute temperature. Perform your calculation at room temperature and at 400K. Compare this minimum energy to a hypothetical process for recovering lithium. Assume that the lithium is obtained through a process that requires evaporating the seawater at near ambient pressure (boiling at near 400K). Calculate the energy required to heat the water from room temperature to 400K (assume that the heat capacity of water is 1 cal/g °K, 4.18 J/g °K, or 0.04 BTU/g °K) and to evaporate the water (assume that the heat of evaporation is 2270 J/g or 2.16 BTU/g).

c. If a gallon of fuel costs $3.00 and each gallon contains approximately 124,000 BTU, what are the energy costs to recover 1 kg of metal at room temperature and at 400K?

Solution:

a. Calculate the change in entropy per mole of mixture:

$$\Delta S = -R \, \Sigma \, x_i \ln x_i = 2 \text{ cal/(°K mole)} * 0.9999998 \ln (0.9999998)$$
$$+ \, 0.0000002 \ln(0.0000002) = -3 * 10^{-6} \text{ cal/(°K mole)}$$

b. Noting that there are roughly $10^9/18$ gram moles per million kilograms of solution (per 0.2 kg of metal), the change in entropy per kilogram of metal is $5*(10^9/18)* -3 * 10^{-6}$ cal/(°K mole) $= -833$ cal/°K. At 300K the energy required to overcome this entropy change, per kilogram of lithium recovered, is 250 kcal; at 400K the energy required is 330 kcal.

c. At room temperature, the change in free energy is roughly 250 kcal per kilogram of metal. At 400K, the change is roughly 330 kcal per kilogram. Both quantities are negligible compared to the energy required to heat 5 million kg of water to 400K (1 cal/g °K * (400–300)°K * $5 * 10^9$ g $= 5 * 10^{11}$ cal $= 5 * 10^8$ kcal). (Note that this does not include the heat of vaporization.)

d. The energy cost for heating 5 million kg of water to 400K is roughly $50,000 ($5 * 10^{11}$ cal * 1 BTU/252 cal * $3/124,000 BTU = $48,000). While boiling seawater is not the only way to recover lithium, this simple example illustrates the difference between theoretical energy requirements and the actual energy requirements of material recovery systems.

The preceding examples demonstrate that the total amounts of resources available in the Earth's crust are sufficient to supply commercial needs indefinitely. However, the amounts of various elements in the Earth's crust that are present in high enough concentrations to be recoverable are limited. The primary limitations on recovering materials are economic. Energy and other resources are required to extract and purify materials, introducing costs. The costs are defined primarily not by physical laws (overcoming entropy) but by the technologies that are used, which continue to evolve.

The amounts of materials that can be economically recovered, referred to as *reserves*, are defined in multiple ways. Figure 4-4 summarizes these definitions. Reserves that are known and that have had their extent demonstrated, and that can be recovered at costs lower than current prices, are referred to as *economically recoverable reserves*. Demonstrated reserves that could be cost-effectively recovered, through either moderate improvements in technology or increases in price, are referred to as *marginal reserves*. Demonstrated reserves that are unlikely to be recoverable at any foreseeable price are referred to as *subeconomic*. Finally, demonstrated reserves can also lead to inferences that similar geological formations may contain similar reserves. These are referred to as *inferred reserves*, which can be economic, marginal, or subeconomic.

Demonstrated and inferred reserves may also be expanded as new areas are opened to exploration, so undiscovered reserves could expand all of the identified resources. Figure 4-4 shows the interrelationships between these ways of referring to reserves.

Example 4-4   McElvey diagram for gold

Using the mineral commodity summaries available from the U.S. Geological Survey (http://minerals.usgs.gov/minerals), identify the reserves for gold in the United States.

Solution: "An assessment of U.S. gold resources indicated 33,000 tons of gold in identified (15,000 tons) and undiscovered (18,000 tons) resources" (USGS, 2010).

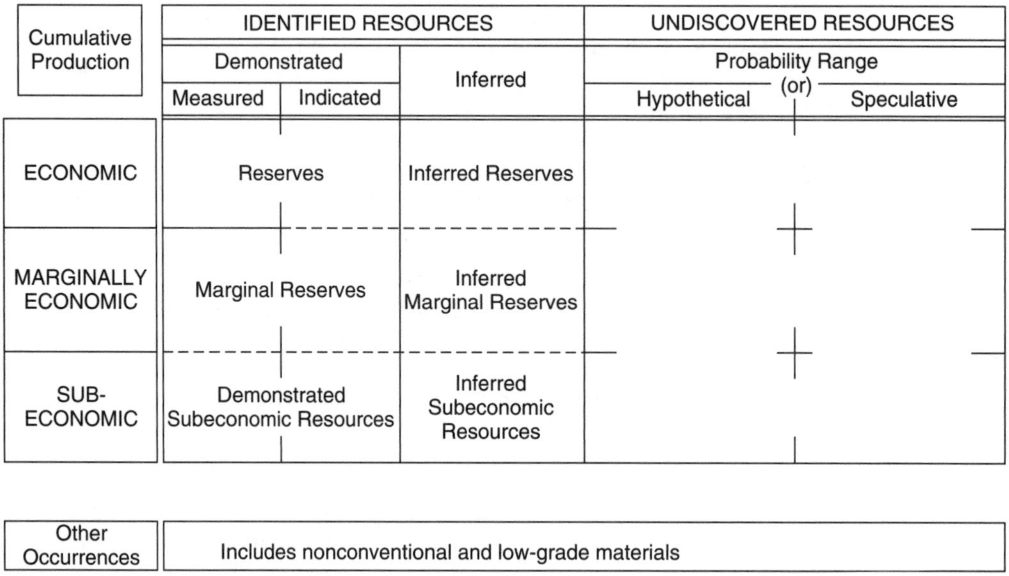

Cumulative Production	IDENTIFIED RESOURCES			UNDISCOVERED RESOURCES		
	Demonstrated		Inferred	Probability Range (or)		
	Measured	Indicated		Hypothetical		Speculative
ECONOMIC	Reserves		Inferred Reserves			
MARGINALLY ECONOMIC	Marginal Reserves		Inferred Marginal Reserves			
SUB-ECONOMIC	Demonstrated Subeconomic Resources		Inferred Subeconomic Resources			

Other Occurrences	Includes nonconventional and low-grade materials

Figure 4-4   A reserve classification for minerals, the McElvey diagram (USGS, 2010, Appendix C)

Example 4-5   McElvey diagram for rare earth elements

Using the mineral commodity summaries available from the U.S. Geological Survey (http://minerals.usgs.gov/minerals), identify the U.S. production and global reserves for rare earth elements in the United States. If an average electric vehicle or plug-in hybrid electric vehicle requires 10 kg more rare earth elements than a conventional vehicle, by what percentage would rare earth metal use increase in the United States if 1 million of these vehicles were manufactured in the United States each year?

Solution: U.S. consumption of rare earth elements was roughly 5000 to 10,000 metric tons per year over the past five years. Almost all of this domestic use, and global use, comes from production in China. Undiscovered resources are believed to be extensive, resulting in an estimated total reserve of 99 million metric tons (USGS, 2010).

One million hybrid or electric vehicles would require roughly 10,000 metric tons (10 million kg) of rare earth elements, roughly doubling U.S. consumption.

To summarize, the total amounts of resources available in the Earth's crust are sufficient to supply commercial needs indefinitely. However, economic reserves are limited. Graedel and Allenby (1995) have evaluated the relative supplies of various elements and grouped them into the categories shown in Table 4-2. Multiple elements are grouped together in the table, since they are found in the same types of ores. For example, Cu ores typically also contain As, Se, and Te. Pt ores typically contain Ir, Os, Pa, Rh, and Ru. The element with the highest demand, or highest price, tends to drive the extraction of these ores. So, for example, mining of Zn

Table 4-2 Supplies of the Elements

Extent of Supply	Elements
Infinite supply	A, Br, Ca, Cl, Kr, Mg, N, Na, Ne, O, Rn, Si, Xe
Ample supply	Al, C, Fe, H, K, S, Ti
Adequate supply	I, Li, P, Rb, Sr
Potentially limited supply	Co, Cr, Mo, Ni, Pb, Pt ores
Potentially highly limited supply	Ag, Au, Cu ores, He, Hg, Sn, Zn ores

Source: Graedel and Allenby, 1995

produces Cd as a by-product, making it available at a price and a quantity that might not be possible if it were not associated with Zn.

Overall, the key concept to take away from this analysis is that the total amount of materials in the crust is enormous. Since, in general, very little of this mass ever escapes the planet, the total amount of material is conserved and these materials will always be available. What does happen as resources are used, however, is that highly concentrated forms of various elements, extractable at low cost, are depleted. While the analyses presented in this section have focused on minerals and metals, the principles are generally valid.

As ever more dilute materials need to be extracted to satisfy demand, the energy and other resources required to perform that extraction, and consequently the cost of the material, increase. This suggests that a logical approach to the efficient use of materials would be to track the flows of materials. If materials, in concentrated form, can be recycled or reused, there is the potential to avoid additional extraction and to save energy. Tracking material flows in engineered systems is the topic of the next section.

4.3   TRACKING MATERIAL FLOWS IN ENGINEERED SYSTEMS

Materials have a life cycle. They are extracted from the lithosphere or biosphere, processed into commodity materials, then products, then used and possibly reused, then eventually disposed of or leaked into the environment. The number of times that a material is reused or recycled before it is released into the environment can have a significant impact on its environmental footprint. In addition, different types of uses, for the same material, can lead to very different types of impacts.

Characterizing the flows and emissions of materials in manufacturing and use requires data on material and mineral flows entering the economy, and information on the wastes, emissions, and recycling structures. Data that enable this new generation of analyses are just emerging. Therefore, terminology and data analysis frameworks are still evolving. One set of terminology is shown in Figure 4-5.

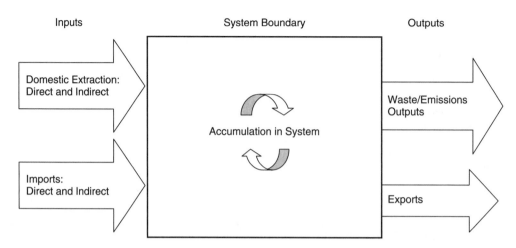

Figure 4-5   Conceptual framework for analyzing material flows (National Research Council, 2003)

As shown in Figure 4-5, material flow analyses are performed on systems with well-defined boundaries. The system boundary might be the geopolitical boundaries of a nation, the natural boundaries of a river's drainage basin, or the technological boundaries of a cluster of industries. The labels for the system inputs and outputs used in Figure 4-5 suggest that the system is a nation, but these inputs and outputs (domestic extraction, imports, and exports) could be labeled "feedstocks" and "products," and the system would then appear to be a cluster of industries.

Figure 4-5 also identifies material inputs, material outputs, releases to the environment, and material accumulation as components of the material flow analysis. The system inputs and outputs are segregated into direct and indirect flows. Direct flows are those that are normally accounted for in engineering analyses, such as fuels, minerals, metals, and water. The indirect, or hidden, flows shown in Figure 4-5 are composed of materials such as mining overburdens and soil erosion from agricultural operations. These hidden material flows do not enter the economic system, yet they occur as a result of economic activity. Within the system, stocks are accumulated and materials are reused and recycled. As flows internal to the system are reengineered to incorporate more reuse and recycling, releases to air, water, and land can be reduced and demands for inputs are reduced.

Consider, as an example of material flow analyses, the element lead (Pb). Pb is a neurotoxin, and Pb exposure is associated with developmental delays (www.epa.gov/iris). Human exposure to Pb should therefore be minimized. Historically, some of the principal uses of lead have been as an octane enhancer in gasoline, in batteries, and in paint. The material flows of Pb in the United States in 1970 and in the mid-1990s are shown in Figure 4-6. As shown in the left-hand portion of the diagram, flows of Pb come from both extraction of geological resources (virgin materials) and recycled material. In 1970 the fraction of virgin material was 36%

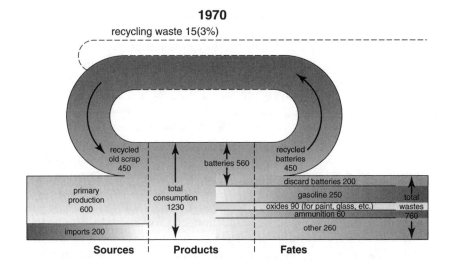

1970

recycling waste 15(3%)

recycled old scrap 450

batteries 560

recycled batteries 450

primary production 600

total consumption 1230

discard batteries 200
gasoline 250
oxides 90 (for paint, glass, etc.)
ammunition 60
other 260

total wastes 760

imports 200

Sources | **Products** | **Fates**

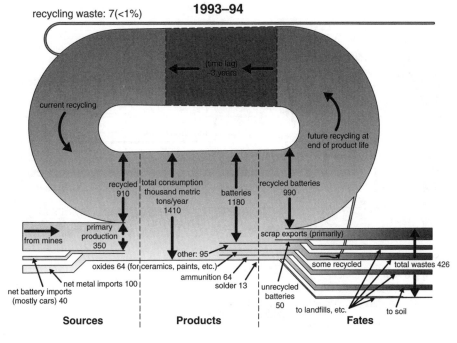

1993–94

recycling waste: 7(<1%)

(time lag) ~3 years

current recycling

future recycling at end of product life

recycled 910

total consumption thousand metric tons/year 1410

batteries 1180

recycled batteries 990

primary production 350

from mines

other: 95

oxides 64 (for ceramics, paints, etc.)

ammunition 64

solder 13

net metal imports 100

net battery imports (mostly cars) 40

scrap exports (primarily)

some recycled

total wastes 426

unrecycled batteries 50

to landfills, etc.

to soil

Sources | **Products** | **Fates**

Figure 4-6 Material cycles for lead in 1970 and the mid-1990s (USGS, 2000); flows are in thousands of metric tons per year

(450 tons recycled/1250 tons total usage), while in the mid-1990s the fraction had increased to 65% (910 tons recycled/1400 tons total usage). The Pb is incorporated into a variety of products; some products, when used as designed (such as lead paint applied outdoors or lead additives in gasoline), result in the release of Pb into the environment (dissipative uses). Other products, when used as designed (lead acid batteries), can be effectively recovered and recycled at the end of the product's life (nondissipative uses).

Figure 4-6 shows that the fraction of dissipative uses decreased significantly from 1970 until the mid-1990s in part because of the phaseout of many lead-based paints and lead additives in gasoline. Use of Pb in batteries increased significantly from 1970 through the mid-1990s, as the demand for starting-lighting-ignition batteries in vehicles increased.

Figure 4-6 can be expanded to incorporate additional types of flows, and Figure 4-7 illustrates some of those flows, using global use of iron as a case study. Figure 4-7 separates iron flows during production of products into production, fabrication/ manufacturing, use, and waste management. There are flows between these stages of materials processing, for example, as "home" scrap within production operations is

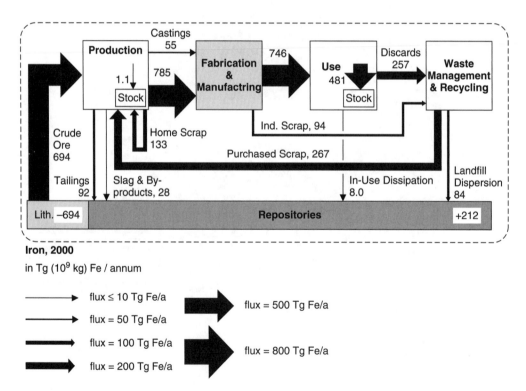

Figure 4-7   Global flows of iron in 2000 (Wang et al., 2007; reprinted with permission, STAF Project, Yale University)

reprocessed. The material flow mapping of Figure 4-6 also shows flows of material from different processing stages into anthropogenic repositories (such as landfills) and the flows of material into durable goods (stock). Overall, however, the mapping provides much of the same information as was illustrated in Figure 4-6. Flows of crude ore from the lithosphere (virgin ore) can be compared to various types of recycling; losses to the environment can be compared to total use and stocks.

Example 4-6   Global material balance for iron

Material flow information, of the type shown in Figure 4-7, typically comes from multiple sources, and material balances can be used to assess the consistency of the information. At the most macroscopic level, the basic material balance equation, in = out + accumulation, can be translated into crude ore in = flows to repositories + stock accumulation. Determine whether the flows in Figure 4-7 satisfy a material balance.

Solution:

$$\text{crude ore in} = \text{flows to repositories} + \text{stock accumulation}$$
$$\text{crude ore in} = 694 \text{ Tg/annum}$$
$$\text{Flows into repositories} + \text{stock accumulation} = 212 + 481 = 693 \text{ Tg/annum}$$

So, within rounding error, the flows satisfy a material balance.

While Figure 4-7 characterizes the flows of iron on a global basis, flows of materials can vary among regions. Mapping regional flows requires the addition of flows from other regions and to other regions (exports and imports); these flows are not required for a global material balance. As an example, Figure 4-8 compares the flows of iron in the United States and China. The United States uses a larger fraction of recycled iron than China.

Understanding the sourcing of materials, and the geopolitical issues associated with source regions, will become increasingly important as global demands on resources increase. Figure 4-9 shows the extent to which various countries are net importers or exporters of iron. The world's leading exporter of iron is Brazil, and the leading importer (for both total imports and net imports [imports-exports]) is the United States. Countries such as Belgium, the Netherlands, and the United Kingdom are both importers of iron as a raw material and exporters of iron in manufactured goods.

The case of iron is not unusual. Global trade in commodity materials leads to some countries becoming net importers, while others are net exporters, depending on the material. For example, while China is a net importer of iron, it is a dominant producer and exporter of rare earth metals.

Combinations of materials scarcity, potential energy savings, and geopolitical factors may lead to increased rates of recycling and reuse for many materials. A qualitative indicator of the potential for such recycling to occur is once again the Sherwood diagram. In Figure 4-3, dilution in virgin ores was shown to be a reasonable predictor of price for many materials. Another way in which

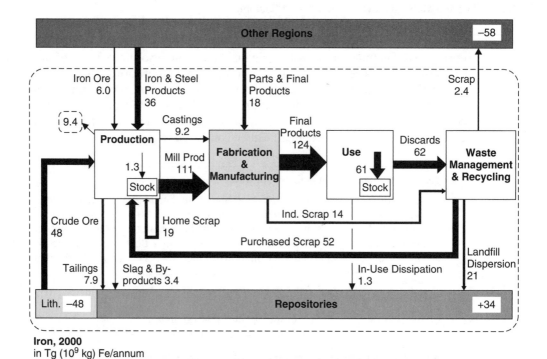

Iron, 2000
in Tg (10^9 kg) Fe/annum

Iron, 2000
in Tg (10^9 kg) Fe/annum

Figure 4-8 Flows of iron in the United States and China, 2000 (Wang et al., 2007; reprinted with permission, STAF Project, Yale University)

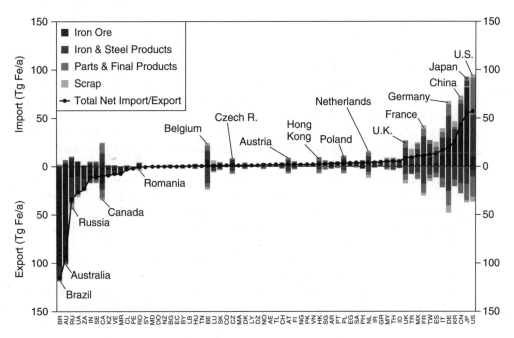

Figure 4-9 Importing and exporting of iron, by country, ranked by net imports minus exports (Reprinted with permission from Wang et al., 2007. Copyright 2007, American Chemical Society)

Table 4-3 Percentage of Metals in Hazardous Wastes in the United States That Can Be Recovered Economically

Metal	Percent Theoretically Recoverable	Percent Recycled in 1986
Sb	74–87	32
As	98–99	3
Ba	95–98	4
Be	54–84	31
Cd	82–97	7
Cr	68–89	8
Cu	85–92	10
Pb	84–95	56
Hg	99	41
Ni	100	0.1
Se	93–95	16
Ag	99–100	1
Tl	97–99	1
V	74–98	1
Zn	96–98	13

Source: Allen and Behmanesh, 1994

this diagram can be used is to assess the degree to which waste streams might be "mined." Allen and Behmanesh (1994) examined the extent to which hazardous wastes in the United States might be mined (cost-effectively recycled) by comparing the degree of dilution of metals in hazardous waste streams to commodity prices. Their original analysis, summarized in Table 4-3, found that many hazardous waste streams in the United States had high enough concentrations of metals to merit additional recycling. Hazardous wastes were chosen because detailed data existed on their compositions, flow rates, and fates. Surprisingly, many hazardous waste streams contain relatively high concentrations of metals. Approximately 90% of the copper and 95% of the zinc found in hazardous wastes were, at the time, at concentrations high enough to recover. For every metal for which data existed, recovery occurred at rates well below rates that would be expected to be economically viable.

This very focused analysis, which was initially performed in 1994 (Allen and Behmanesh, 1994), led to the conclusion that many opportunities existed for recovering materials from wastes. There are limitations to the analysis, however. The analysis focused only on hazardous wastes, where legal liability concerns may limit the desire to recycle. The identification of "recyclable" streams was simplistic. It ignored issues related to economies of scale (i.e., processing geographically dispersed, heterogeneous waste streams may be more expensive than extracting a relatively homogeneous ore from a single mine). Nevertheless, the analysis indicated that resources are not effectively recovered from many waste streams.

The United Nations has assembled more recent, global data on metals recycling. Its analysis continues to conclude that global recycling rates can continue to be improved. Metals like Pb (along with Fe, Cr, Co, Ni, Cu, Zn, and many precious metals) have postconsumer recycling rates that exceed 50% globally, but many other metals (e.g., rare earth elements) have postconsumer recycling rates of less than 1% (United Nations Environment Programme, 2011).

One of the primary barriers to using wastes as raw materials is a lack of critical information on waste streams. While a large number of data sources are available on waste streams, they lack critical information that is needed to assess whether the waste streams might be reused. Data on the composition of wastes, their location, and co-contaminants are rarely available yet are critical to evaluating the potential use of wastes as raw materials.

Example 4-7   Energy savings due to recycling

Aluminum is extensively recycled, in part because of the energy savings associated with reprocessing recycled aluminum into aluminum stock, as compared to the processing of bauxite into aluminum. The table below shows the differences in fuel and electricity inputs required for primary aluminum (from bauxite) and aluminum recycled from automotive scrap (U.S. Life Cycle Inventory data available at www.nrel.gov/lci). Using current prices for the fuels and electricity, calculate the differences in energy costs, per pound of aluminum. How does this compare to the market price of aluminum?

	Energy Inputs Required for 1000 Pounds of Primary Aluminum Ingot Production from Bauxite	Energy Inputs Required for 1000 Pounds of Recycled Aluminum Ingot Production from Automotive Scrap
Total electricity (ingot formation)	7240 kWh	303 kWh
Coal	16.6 lb	
Coke	0.0026 lb	
Distillate oil	2.98 gal	
Gasoline	0.20 gal	
Natural gas	8,606 ft^3	3570 cu ft
Propane/LPG	0.84 gal	
Residual oil	26.8 gal	

Solution: Energy inputs are dominated by liquid fuels, electricity, and natural gas use. At a cost of $3.00 per gallon of liquid fuel, $0.10 per kilowatt-hour, and $5.00 per thousand standard cubic feet (tscf) of natural gas, the energy cost per 1000 lb of primary aluminum is $860 (30 gal * $3.00/gal + 7240 kWh * $0.10/kWh + 8600 scf * $5.00/tscf). In contrast, the energy cost per 1000 lb of recycled aluminum is $48 (303 kWh * $0.10/kWh + 3570 scf * $5.00/tscf). This is a cost savings of roughly $0.81 per pound of aluminum. In contrast, the price of aluminum on commodity markets (2010) is $1.10 to $1.20 per pound.

To summarize, material flow tracking can identify overall availability of materials and potential opportunities for material reuse or recycling. One way to view these flows is as an opportunity to change the designs of industrial systems so that they more closely resemble highly networked, mass-conserving, natural ecosystems, an industrial ecology. As this text is being written, the data necessary to perform detailed material tracking are just beginning to emerge in a consistent framework. For many analyses, it will be necessary to assemble the necessary data on material use and flows on a case-specific basis. Nevertheless, these types of analyses will become increasingly valuable as material scarcity becomes an important issue.

4.4   ENVIRONMENTAL RELEASES

A material, at the end of its useful life cycle or as a consequence of its production or use, is released into the environment. Will its release pose significant environmental or human health risks? Will the chemical degrade in the environment or will it persist?

The challenges involved in answering these questions are formidable. Tens of thousands of chemicals are produced commercially, and every year, a thousand or more new chemicals are developed and introduced into commerce. For any chemical in use, there are a number of potential risks to human health and the environment.

In general, it is not practicable to rigorously and precisely evaluate all possible environmental impacts. Nevertheless, a preliminary screening of the potential environmental impacts of chemicals is necessary and is possible. Preliminary risk screenings allow businesses, government agencies, and the public to identify problem chemicals and to identify potential risk reduction opportunities. The challenge is to perform these preliminary risk screenings with a limited amount of information.

The goal of this section is to describe qualitative and quantitative methods for estimating environmental risks when the only information available is a chemical structure. Many of these methods have been developed by the U.S. EPA and its contractors. The methods are routinely used in evaluating Premanufacture Notices (PMNs) submitted under the Toxic Substances Control Act (TSCA). Under the provisions of TSCA, before a new chemical can be manufactured in the United States, a PMN must be submitted to the EPA. The notice specifies the chemical to be manufactured, the quantity to be manufactured, and any known environmental impacts as well as potential releases from the manufacturing site. Based on these limited data, the EPA must assess whether the manufacture or use of the proposed chemical may pose an unreasonable risk to human or ecological health. To accomplish that assessment, a set of tools has been developed that relate chemical structure to potential environmental risks.

Table 4-4 identifies the chemical and physical properties that will influence the processes that determine environmental exposure and hazard. The table makes clear that a wide range of properties needs to be estimated to perform a screening-level assessment of environmental risks.

The first group of properties that must be estimated in an assessment of environmental risk are the basic physical and chemical properties that describe a chemical's partitioning among solid, liquid, and gas phases. These properties are important in determining whether a pollutant will concentrate in air, water, soil, or living organisms and include melting point, boiling point, vapor pressure, and water solubility. Additional molecular properties, related to phase partitioning, that are frequently used in assessing the environmental fate of chemicals include the

Table 4-4 Chemical Properties Needed to Perform Environmental Risk Screenings

Environmental Process	Relevant Properties
Dispersion and fate	Volatility, density, melting point, water solubility, effectiveness of wastewater treatment
Persistence in the environment	Atmospheric oxidation rate, aqueous hydrolysis rate, photolysis rate, rate of microbial degradation, and adsorption
Uptake by organisms	Volatility, lipophilicity, molecular size, degradation rate in organism
Human uptake	Transport across dermal layers, transport rates across lung membrane, degradation rates within the human body
Toxicity and other health effects	Dose-response relationships

octanol-water partition coefficient, soil sorption coefficients, Henry's law constants, and bioconcentration factors. (Each of these properties is defined in Table 4-5.) Once the basic physical and chemical properties are defined, a series of properties that influence the persistence of chemicals in the environment are estimated. These include estimates of the rates at which chemicals will react in the atmosphere, the rates of reaction in aqueous environments, and the rates at which the compounds will be metabolized by organisms. If environmental concentrations can be estimated based on release rates and environmental fate and persistence properties, human exposures to the chemicals can be estimated. Finally, if exposures and hazards are known, risks to humans and the environment can be estimated.

These chemical and physical properties can be used to evaluate a variety of metrics related to environmental impacts; some of the most commonly evaluated environmental metrics are persistence, bioaccumulation, and toxicity. For any one of these metrics of environmental impact, it may be necessary to consider a number of properties in performing an evaluation. For example, in evaluating persistence, atmospheric half-lives and biodegradation half-lives may be needed. In evaluating toxicity, it may be necessary to consider a variety of ecotoxicity measures and human toxicity measures. Because there is such a wide variety of criteria that can be used in evaluating environmental risks—ranging from human carcinogenicity to biodiversity—and because opinions vary widely on the relative importance of the evaluation criteria, there is no single evaluation methodology that is universally accepted for evaluating the environmental hazards of chemicals.

While it is not the only approach, measures of persistence, bioaccumulation, and toxicity are used by the EPA in evaluating PMNs under TSCA, and in performing these evaluations, the EPA classifies chemicals into categories of high, moderate, and low concern. Methods for estimating persistence, bioaccumulation, and toxicity rely on estimates of physical and chemical properties, such as those listed in Tables 4-4 and 4-5. Methods for estimating these properties are described by Allen and Shonnard (2001), and the impact of the properties on assessing environmental persistence and fate is presented in problems at the end of this chapter.

4.4.1 Using Property Estimates to Evaluate Environmental Partitioning, Persistence, and Measures of Exposure

This section will briefly illustrate, through examples, how chemical and physical properties can be employed to estimate environmental partitioning, persistence, and measures of exposure. The problems associated with estimating environmental exposures are complex. Consider the relatively simple example of calculating exposure through drinking contaminated surface water. Assume that a chemical is released to a river upstream of the intake to a public drinking water treatment plant. To evaluate the exposure we would need to determine

- What fraction of the chemical was adsorbed by river sediments
- What fraction of the chemical was volatilized to the atmosphere
- What fraction of the chemical was taken up by living organisms

Table 4-5   Properties That Influence Environmental Phase Partitioning

Property	Definition	Significance in Estimating Environmental Fate and Risks
Melting point (T_m)	Temperature at which solid and liquid coexist at equilibrium	Sometimes used as a correlating parameter in estimating other properties for compounds that are solids at ambient or near-ambient conditions
Boiling point (T_b)	Temperature at which the vapor pressure of a compound equals atmospheric pressure; normal boiling points (temperatures at which pressure equals one atmosphere) will be used in this text	Characterizes the partitioning between gas and liquid phases; frequently used as a correlating variable in estimating other properties
Vapor pressure (P_{vp})	Partial pressure exerted by a vapor when the vapor is in equilibrium with its liquid	Characterizes the partitioning between gas and liquid phases
Henry's law constant (H)	Equilibrium ratio of the concentration of a compound in the gas phase to the concentration of the compound in a dilute aqueous solution (sometimes reported as atm-m^3/mol; the dimensionless form will be used in this text); $C_A = H\,C_W$	Characterizes the equilibrium partitioning between gas and aqueous phases
Octanol-water partition coefficient (K_{OW})	Equilibrium ratio of the concentration of a compound in octanol to the concentration of the compound in water; $C_O = K_{OW}\,C_W$	Characterizes the partitioning between hydrophilic and hydrophobic phases in the environment and the human body; frequently used as a correlating variable in estimating other properties
Water solubility (S)	Equilibrium solubility in mol/L	Characterizes the partitioning between hydrophilic and hydrophobic phases in the environment
Soil sorption coefficient (K_{OC})	Equilibrium ratio of the mass of a compound adsorbed per unit weight of organic carbon in soil (in µg/g organic carbon) to the concentration of the compound in the water phase (in µg/ml); $C_S = K_{OC}\,C_W$	Characterizes the partitioning between solid and liquid phases in soil which in turn determines mobility in soils; frequently estimated based on the octanol-water partition coefficient, and water solubility
Bioconcentration factor (BCF)	Ratio of a chemical's concentration in the tissue of an aquatic organism (C_{AO}) to its concentration in water (reported as L/kg); $C_{AO} = BCF\,C_W$	Characterizes the magnification of concentrations through the food chain

- What fraction of the chemical was biodegraded or was lost through other reactions
- What fraction of the chemical was removed by the treatment processes in the public water system

In this case, exposure estimates will require information on the soil sorption coefficient, vapor pressure, water solubility, bioconcentration factor, and biodegradability of the compound, as well as river flow rates, surface area, sediment concentration, and other parameters. Often, for screening calculations, equilibrium partitioning of compounds is assumed between water and soil, water and air, and water and living systems. Assuming equilibrium partitioning allows ratios of concentrations to be determined. For example, the Henry's law coefficient defines the equilibrium ratio of air and water concentrations. The soil sorption coefficient allows equilibrium ratios of soil and water concentrations to be determined, and bioconcentration factors determine the ratio of biota and water concentrations. A simple, yet typical, set of calculations is shown in Example 4-8.

Example 4-8   Environmental Partitioning

Assume that a chemical with a molecular weight of 150 is released at a rate of 300 kg/day to a river, 100 km upstream of the intake to a public water system. Estimate the initial equilibrium partitioning of the chemical in the water, sediment, and biota.

Data: Soil sorption coefficient: 10,000 (ratio of concentration of pollutant in soil to the concentration in water)
Organic solids concentration in suspended solids: 15 ppm
River flow rate: 500 million L per day
Bioconcentration factor: 100,000 (ratio of concentration of pollutant in biota (fish) to the concentration in water)
Biota loading: 100 g per 100 cubic meters

Solution: The ratio of concentrations in water, sediment, and biota will be approximately

$$1 : 10,000 : 100,000$$

Based on the river flow rate, the total flow rates of water, sediment, and biota are

Water: (500 million L/day * 1 kg/L) = 500 million kg/day
Sediment: 500 million kg/day * 15 kg sediment/million kg water = 7500 kg sediment/day
Biota: 500 million kg/day * 0.1 kg biota/million kg water = 50 kg biota/day

Performing a mass balance:

$$300 \text{ kg/day} = 500 \text{ million kg water/day } (C_{water}) + 7500 \text{ kg sediment/day}$$
$$(10,000 \ C_{water}) + 50 \text{ kg biota/day } (100,000 \ C_{water})$$

where (C_{water}) is the chemical concentration in the water phase,

$$(C_{water}) = 0.5 * 10^{-6} \text{ kg chemical/kg water} = 0.5 \text{ ppm}$$
$$\text{Concentration in sediment} = 0.5 \text{ ppm} * 10,000 = 5000 \text{ ppm}$$

$$\text{Concentration in biota} = 0.5 \text{ ppm} * 100{,}000 = 50{,}000 \text{ ppm}$$

The ratio of the mass in water, sediment, and biota is

$$500{,}000{,}000 : 7500 \ (5000) : 50 \ (50{,}000)$$
$$84 : 13 : 1$$

Thus, although the concentrations are much higher in the biota and the sediment phases, more than 80% of the mass remains in the water phase.

4.4.2 Direct Use of Properties to Categorize the Environmental Risks of Chemicals

Section 4.4.1 illustrated how chemical and physical properties can be used to estimate environmental footprints. Another approach is to use the values of properties to directly categorize the environmental risks of chemicals. For example, Table 4-6 is a summary of the categories used by the EPA to classify the persistence and bioaccumulation of chemicals.

For each of these categories, a score might be given. For example, persistence might be scored 1 through 4 for the four levels of biodegradation listed in Table 4-6. Bioaccumulation might be given a score of 1 through 3 based on the three categories of bioconcentration factor. Similar scores could be developed for toxicity. These indices or scores could then be combined, for example, by adding the scores, to arrive at a composite index.

Other categories could be used to classify materials. Example 4-9 compares a variety of chemicals used in fuels.

Example 4-9   Classifying fuel molecules

Compare the soil sorption, water solubility, and biodegradation of three compounds that have been used in gasoline: ethanol, methyl-tert butyl ether (MTBE), and isohexane. Use these data to assess which of the chemicals, if spilled on land, would be more likely to migrate to surface or groundwater. Isohexane is one of the most commonly found molecules in gasoline derived from petroleum. Ethanol is commonly obtained from corn grain, and MTBE was produced in large quantities in the United States from methanol (derived from natural gas) and light petroleum gases (isobutylene).

Table 4-6　Classification Criteria for Persistence and Bioaccumulation

Property	Classifications
Water Solubility (S)	
Very soluble	S > 10,000 ppm
Soluble	1,000 < S < 10,000 ppm
Moderately soluble	100 < S < 1,000 ppm
Slightly soluble	0.1 < S < 100 ppm
Insoluble	S < 0.1 ppm
Soil Sorption	
Very strong sorption	$Log\ K_{oc} > 4.5$
Strong sorption	$4.5 > Log\ K_{oc} > 3.5$
Moderate sorption	$3.5 > Log\ K_{oc} > 2.5$
Low sorption	$2.5 > Log\ K_{oc} > 1.5$
Negligible sorption	$1.5 > Log\ K_{oc}$
Biodegradation	
Rapid	>60% degradation over 1 week
Moderate	>30% degradation over 28 days
Slow	<30% degradation over 28 days
Very slow	<30% degradation over more than 28 days
Bioaccumulation Potential	
High potential	$8.0 > Log\ K_{oc} > 4.3$ or BCF > 1000
Moderate potential	$4.3 > Log\ K_{ow} > 3.5$ or 1000 > BCF > 250
Low potential	$3.5 > Log\ K_{oc}$ or 250 > BCF

Solution:　Properties for these compounds can be determined through property databases or using methods described in Allen and Shonnard (2001).

Chemical (CAS Registry Number)	Soil Sorption (Log K_{OC})	Water Solubility (ppm)	Biodegradation (Half-life)
Ethanol (64-17-5)	0.3	Infinite	Days–weeks
MTBE (1634-04-4)	1.3	40,000	Days–weeks
Isohexane (107-83-5)	2.0	14	Days–weeks

The results indicate that petroleum-derived gasoline, if spilled onto ground, is more likely to adsorb to soils than MTBE and ethanol, making it less likely to migrate to water sources and cause water contamination. In contrast, however, MTBE and ethanol have been added to gasoline because of provisions of the Clean Air Act that are designed to reduce emissions of air pollutants. So, while ethanol and MTBE are more likely to find their way into surface and groundwaters than components of conventional gasoline, their use can improve air quality.

4.5   SUMMARY

Almost every engineering design involves the use of materials, and these materials have environmental footprints. Methods for characterizing the footprints associated with extraction, processing, and environmental releases of materials have been described in this chapter, but these assessments can lead to very different characterizations of materials. Is the material scarce? Can it be recycled? Do environmental releases have significant impacts?

There are no universally accepted methods for combining these characterizations of whether a material is sustainable. Multiple methods are used. Engineers will need to consider materials in the context of particular designs, recognizing that the choice of the most sustainable materials will be application-specific.

PROBLEMS

1. **Material use in photovoltaic manufacturing** Approximately 65 gigawatts of electrical power are consumed during periods of peak demand in Texas. If the entire demand were to be met by copper-indium-gallium-diselenide (CIGS) photovoltaic cells requiring 50 kg of indium and 10 kg of gallium per megawatt of power generated, calculate the total amounts of Ga and In required to manufacture these cells, and compare these quantities to the total worldwide production of Ga and In (see U.S. DOE, 2010).

2. **Materials use in wind turbine manufacturing** Approximately 65 gigawatts of electrical power are consumed during periods of peak demand in Texas. If the entire demand were to be met by wind power, and the wind turbines require 25 kg of dysprosium per megawatt of power generated, calculate the total amount of Dy required to manufacture these turbines, and compare this quantity to the total worldwide production of Dy (see U.S. DOE, 2010).

3. **Recycling rare earth elements from hybrid vehicles** Estimate the potential flow of recycled rare earth elements that could be recovered from plug-in hybrid electric vehicles (PHEVs) by 2030. Assume that 10 kg of rare earth elements are used in a typical PHEV and that, by 2030, 5 million vehicles per year are retired and available for recycling. Compare the potential flow of recycled material to current mine production.

4. **Ranking critical materials** The U.S. Department of Energy has used a method for categorizing critical materials that involves plotting the ranking of both the supply and the potential demand, where demand includes assessments of both the growth in markets for products and the development of alternative materials. Select one of the materials from the Department of Energy assessment and write a one-page summary describing the basis for the rankings (U.S. DOE, 2010).

5. **Materials use in green products** The EPA has developed a Design for Environment labeling program for a variety of products. Visit the Web site for the program and write a one-page summary of the criteria used in awarding the Design for Environment label (www.epa.gov/oppt/dfe/index.htm or www.epa.gov/oppt/dfe/pubs/projects/gfcp/index.htm#Standard).

6. **Environmental fate of gasoline substitutes** The compound methyl-tert butyl ether (MTBE) has been used extensively as a gasoline additive. If 100 kg of this compound were accidentally spilled into a lake over the course of a summer, calculate the concentrations in water, sediment, and fish (neglect volatilization) that would result. Assume that the volume of the lake is $8 * 10^7$ m^3, the organic solids loading is 20 ppm, and the fish loading is 2 kg per 10^4 m^3. Assume that the lake is well mixed.
   a. What are the concentrations in water, sediment, and fish?
   b. What fraction of the MTBE would be found in the sediment?
   c. What fraction would be found in the fish?
   d. If you swallowed 1 L of water during a day while waterskiing, how much MTBE would you ingest?
       Soil adsorption coefficient: 11.56 L/kg
       Bioaccumulation factor: 3.162 L/kg wet-wt

7. **Environmental fate of medications** Medications can enter wastewater management systems either through human excretions or through improper disposal. Many medications are not effectively treated by current wastewater management processes and therefore are discharged by wastewater treatment plants. Antidepressants are some of the most commonly prescribed medications in the United States, and many antidepressants are not effectively managed by wastewater treatment plants.
   a. A wastewater treatment plant discharges to Boulder Creek in Colorado at a rate of 64 million L/day, and the stream flow, upstream of the wastewater discharge point, is 1110 L/s. If the concentration of the antidepressant venlafaxine (Effexor) measured in the creek water is 10 ng/L, what is the mass discharged per day from the wastewater treatment plant, assuming that there are no sources other than the wastewater treatment plant, and assuming that the drug rapidly equilibrates among sediment, fish, and water? What is the concentration of the medication in the wastewater treatment plant effluent?

$$\text{BCF} = 40.27$$
$$K_{OC} = 3.162$$
$$\text{Organic sediment} = 15 \text{ ppm}$$
$$\text{Biota concentration} = 5 \text{ g/100 m}^3$$

   b. A man fishing near the outflow point of Boulder Creek eats 0.2 kg of fish from the creek. How much venlafaxine will he ingest? One dose of Effexor contains 75 mg of venlafaxine. What percentage of a dose will the fisherman ingest?

REFERENCES

Allen, D. T. 1996. "Waste Exchanges and Material Recovery." *Pollution Prevention Review* 6(2):105–12.

Allen, D. T., and N. Behmanesh. 1994. "Wastes as Raw Materials. In *The Greening of Industrial Ecosystems*, edited by B. R. Allenby and D. J. Richards. Washington, DC: National Academy Press, pp. 69–89.

Allen, D. T., and D. R. Shonnard. 2002. *Green Engineering: Environmentally Conscious Design of Chemical Processes*. Upper Saddle River, NJ: Prentice Hall.

Gerst, M., and T. E. Graedel. 2008. "In-Use Stocks of Metals: Status and Implications." *Environmental Science & Technology* 42:7038–45.

Graedel, T. E., and B. R. Allenby. 1995. *Industrial Ecology*. Englewood Cliffs, NJ: Prentice Hall.

Johnson, J., E. M. Harper, R. Lifset, and T. E. Graedel. 2007. "Dining at the Periodic Table: Metals Concentrations as They Relate to Recycling." *Environmental Science & Technology* 41:1759–65.

Kesler, S. E. 1994. *Mineral Resources, Economics and the Environment*. New York: Macmillan.

National Research Council. 2003. *Materials Count*. Washington, DC: National Academy Press.

Taylor, S. R. 1964. "Trace Element Abundances and the Chondritic Earth Model." *Geochimica et Cosmochimica Acta* 28(12):1989–98.

United Nations Environment Programme. 2011. *Recycling Rates of Metals: A Status Report*. Available at www.unep.org/resourcepanel/Publications/Recyclingratesofmetals/tabid/56073/Default.aspx. Accessed July 2011.

U.S. DOE (U.S. Department of Energy). 2010. *Critical Materials Strategy*. December. Available at www.energy.gov/news/documents/criticalmaterialsstrategy.pdf.

USGS (United States Geological Survey). 2000. *Materials and Energy Flows in the Earth Science Century*. Circular 1194.

———. 2010. *Mineral Commodities Summaries*. Washington, DC: U.S. Government Printing Office.

Wang, T., D. B. Müller, and T. E. Graedel. 2007. "Forging the Anthropogenic Iron Cycle." *Environmental Science & Technology* 41:5120–29.

CHAPTER 5

Design for Sustainability: Economic, Environmental, and Social Indicators

5.1   INTRODUCTION

Engineers play an important role in global sustainable development by designing production systems for materials, minerals, chemicals, energy, electricity generation and distribution, transportation, buildings and other structures, and consumer products. These designs have impacts on the environment, economies, and societies at spatial scales that vary from local to global and at temporal scales that vary from minutes to decades. As engineers create designs, they not only evaluate their designs at multiple spatial and temporal scales, they also embed their designs in complex systems.

The field of transportation provides an illustration of the multiple layers of systems in which engineers create designs. Among the most visible products designed by engineers are automobiles. Engineers design engines, and improvements to the design of a fossil-fuel-powered engine for an automobile can increase fuel efficiency and reduce the environmental impacts of emissions associated with burning fuels, while simultaneously reducing the cost of operating the vehicle. The size, power, and fuel efficiency of the engine must be balanced with the weight of the vehicle, however, so changes in engine design must be considered within the entire vehicle system. Further reductions in emissions and operating costs might be possible by lowering the weight of the vehicle. The use of materials and fuels by automobiles are embedded in complex fuel and material supply systems. Developing systems to recycle the materials that make up the automobile at the end of its useful life might improve the environmental and economic performance of global material flows. Use of alternative power sources, such as biofuels or electricity, can impact global flows of fuels, which, in turn, might impact global flows of materials such as water. Finally, the design of cities that reduce the need for personal transportation could dramatically reduce the environmental impacts of transportation systems and

117

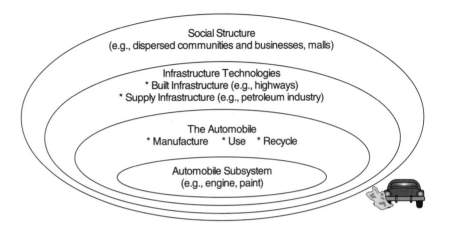

Figure 5-1 Engineering design for sustainability can consider a variety of system scales, as shown for the automobile: gate-to-gate (subsystem), cradle-to-grave (the automobile), interindustry/infrastructure, and extra-industry/societal. (From Graedel and Allenby, *Industrial Ecology and the Automobile*, Copyright 1998. Printed and electronically reproduced by permission of Pearson Education, Inc., Upper Saddle River, New Jersey)

would also transform social structures. These multiple layers of systems in which engineering design decisions are embedded are shown conceptually in Figure 5-1.

The point of this example is to illustrate that sustainable design of engineered systems will lead to consideration of multiple spatial and temporal scales and will require that the engineer interact with professionals with many different backgrounds both within and outside of engineering. These sustainable design challenges are complex, and the tools for addressing these problems are still emerging. This chapter will introduce the tools that are currently available, which are a combination of principles and quantitative analysis methods. The chapter begins with general principles of sustainable engineering design; the principles are illustrated with examples. Next, because sustainable designs must be able to compete in the marketplace, quantitative methods for evaluating, and in some cases monetizing, environmental externalities will be discussed. The use of other, nonmonetized environmental indicators for evaluating process technologies and product alternatives will also be described. Indicators of social impact for engineering design are only at the initial stage of development, yet some prototype methods have been proposed and used, and an introduction to this emerging area will be presented.

5.2   SUSTAINABLE ENGINEERING DESIGN PRINCIPLES

Starting from the introduction of sustainability development concepts in the "Brundtland Report" (WCED, 1987), there have been many attempts to incorporate sustainability principles into engineering design. For example, the "Hannover

Principles" express the view that human systems must be designed to coexist with natural systems, renewable resources should be used, safe and long-lived products are desired, and elimination of waste is a priority (McDonough and Braungart, 1992). The Augsburg Materials Declaration (2002)[1] identifies eight factors that must be considered to achieve sustainable production, including integration of environmentally benign design, materials, and manufacturing over all stages of the life cycle; optimization and exploitation of raw materials and natural resources; energy-efficient production technologies and product distribution; regenerative energy sources; and durability, recyclability, and closed loops. The "12 Principles of Green Engineering" (Anastas and Zimmerman, 2003) include design to be inherently safe and benign, design for recycle or a commercial afterlife, energy and mass efficiency, and integration with existing energy and material flows. The Sandestin Green Engineering Principles, developed as an outcome of a multidisciplinary engineering conference, emphasize the need for holistic thinking and the use of environmental impact and integrative analysis tools such as life-cycle assessment (Abraham and Nguyen, 2003; Shonnard et al. 2007). The Sandestin Green Engineering Principles (also referred to as "Sustainable Engineering Principles"; Abraham, 2006) were developed based on a starting list of principles compiled from a literature review of available sustainability or green-related principles and declaratory statements, including the Hannover Principles, CERES,[2] the Augsburg Materials Declaration, the Twelve Principles of Green Chemistry, Ahwahnee Principles,[3] and Earth Charter Principles.[4] These sustainability concepts and engineering design principles can be summarized in this statement: *The goal of sustainable engineering design is to create products that meet the needs of today in an equitable fashion while maintaining healthy ecosystems and without compromising the ability of future generations to meet their resource needs.*

The **Sandestin Sustainable Engineering Principles** and the **12 Principles of Green Engineering** are illustrative of the multiple sets of principles available. They capture similar, but also complementary, elements of sustainability and engineering design, and therefore these two sets of principles will be described in more detail. Here are the Sandestin Green Engineering Principles (Sustainable Engineering Principles) (Abraham and Nguyen, 2003; Shonnard et al., 2007):

Principle 1: Engineer processes and products holistically, use system analysis, and integrate environmental impact assessment tools. These concepts resonate in a number of Green and Sustainable Engineering principles and are addressed at length in various textbooks, including *Green Engineering: Environmentally Conscious Design of Chemical Processess* by the authors (Allen and Shonnard). The principle points out the importance of systematic evaluation and reduction of human health and environmental impacts of designs, products, technologies, processes, and systems. The use of system-based techniques such as heat and mass integration techniques is

---

1. Available at http://idw-online.de/denews53234 and www.nanoforum.org/dateien/temp/armin%20Reller.pdf? 07062006023521.
2. CERES (Coalition for Environmentally Responsible Economies) Principles, www.ceres.org/our_work/principles.htm.
3. "Ahwahnee Principles," Local Government Commission, www.lgc.org/ahwahnee/principles.html.
4. "Earth Charter Principles," www.earthcharterusa.org/earth_charter.html.

Sandestin Sustainable Engineering Principles

Green Engineering transforms existing engineering disciplines and practices to those that promote sustainability. Green Engineering incorporates development and implementation of technologically and economically viable products, processes and systems that promote human welfare while protecting human health and elevating the protection of the biosphere as a criterion in engineering solutions. To fully implement green engineering solutions, engineers use the following principles:

Principle 1: Engineer processes and products holistically, use system analysis, and integrate environmental impact assessment tools.

Principle 2: Conserve and improve natural ecosystems while protecting human health and well-being.

Principle 3: Use life cycle thinking in all engineering activities.

Principle 4: Ensure that all material and energy inputs and outputs are as inherently safe and benign as possible.

Principle 5: Minimize depletion of natural resources.

Principle 6: Strive to prevent waste.

Principle 7: Develop and apply engineering solutions, while being cognizant of local geography, aspirations, and cultures.

Principle 8: Create engineering solutions beyond current or dominant technologies; improve, innovate and invent technologies to achieve sustainability.

Principle 9: Actively engage communities and stakeholders in development of engineering solutions.

There is a duty to inform society of the practice of Green Engineering.

---

essential to minimize human health and environmental impacts of designs through material and energy optimization. The principle also conveys the importance of not shifting risk (e.g., reducing releases to one environmental medium may increase risk to another medium and/or increase the likelihood of worker exposures and jeopardize worker safety (Shonnard et al. 2007; Abraham, 2006). A well-known example of the consequences of shifting risk is the use of methyl-tert butyl ether (MTBE) as a gasoline additive. MTBE was added to gasoline to reduce emissions of carbon monoxide from automobile tailpipes, thereby protecting human health. Its greater mobility in soil and water environments, however, meant that spills of MTBE could more readily migrate to and disperse in water supplies than spills of gasoline (for more details see www.epa.gov/otaq/consumer/fuels/oxypanel/blueribb.htm).

    Principle 2: Conserve and improve natural ecosystems while protecting human health and well-being. This principle expresses the importance of understanding environmental processes for engineers involved in design of chemicals, automobiles, buildings, and other manufactured goods. There are many examples where a lack of understanding caused severe environmental harm and raised the level of health risk to humans and other forms of life. For example, chlorofluorocarbons (CFCs) were thought to be ideal refrigerants. They replaced dangerous refrigerant fluids like ammonia and made storage of food and building climate control far safer. Their benefits to human health and well-being were clear; however, once the role of CFCs in stratospheric ozone destruction chemistry was worked out, it became clear that there were hazards to human health associated with CFC use. Not every engineer needs to be an expert in environmental processes and health effects, but designers should be aware of the potential harm that can be caused and work with multidisciplinary experts to achieve more sustainable solutions.

Principle 3: Use life-cycle thinking in all engineering activities. This principle complements Principle 1. Every engineered product is created, functions over a useful life, and is eventually disposed of to the environment. Life-cycle thinking can help avoid a narrow outlook on environmental, social, and economic concerns and help make informed decisions. Life-cycle thinking that gets incorporated into design will help identify design alternatives that minimize environmental impacts at the various life stages. This same kind of thinking can also consider economic and societal aspects. The importance of life-cycle thinking can be illustrated (see Chapter 6) through the case study of the greenhouse gas emissions associated with biofuels.

Principle 4: Ensure that all material and energy inputs and outputs are as inherently safe and benign as possible. This principle complements Principle 3. These characteristics of materials and chemicals must be applied to all stages of a product's life, from extraction to use and disposal. The following are some questions that must be asked in each stage of the life cycle: Are the materials toxic? Are there inherently benign (in terms of toxicity) materials that can be used as substitutes? Will exposure during manufacturing be a health problem to workers? Does the product pose minimal impact during recycle and disposal? Will an unintentional release of material quickly degrade in the environment? Properties of materials relevant to safety, beyond toxicity, must also be considered, such as flammability, explosivity, and corrosivity. Chapter 4 provides a number of examples of the selection of materials, based on their environmental and functional properties.

Principle 5: Minimize depletion of natural resources. As the world's population continues to grow and becomes more affluent, natural resources will be used at ever greater rates, and the importance of this principle is raised. An overview of materials, energy, and water use is provided in Chapter 1 of this text. Efficient use of nonrenewable energy resources is of primary interest, and development of renewable alternatives for energy and materials should be given a high priority in engineering design.

Principle 6: Strive to prevent waste. Waste not only represents material that takes up space in landfills but more important represents a loss of efficiency in a production system that includes many input materials and energy sources. When waste is avoided through design, the environmental impacts associated with the input materials and the energy that went into producing the discarded product are also avoided. The following are some questions that must be asked during design: Are there ways (e.g., procedures, engineering) to improve the yield or efficiency of raw materials? How can the releases or wastes be recycled and reused? Can the product be reused after its normal commercial life, hence minimizing the raw materials needed to manufacture new products? Methods for minimizing wastes, particularly in chemical processes, are discussed at length in various textbooks, including *Green Engineering: Environmentally Conscious Design of Chemical Processes* (Allen and Shonnard).

Principle 7: Develop and apply engineering solutions, while being cognizant of local geography, aspirations, and cultures. Engineering designs are directed toward meeting individual human and societal needs, and in order to better achieve this goal, awareness of the societal context of the design is crucial. An engineering design in one society, such as rapid public transportation systems, may not meet the aspirations and needs in another society, even though the design achieves environmental objectives. The main point is to move each society, through engineering design, toward more sustainable utilization of resources in a way that achieves that society's or individual's aspirations.

Principle 8: Create engineering solutions beyond current or dominant technologies; improve, innovate, and invent (technologies) to achieve sustainability. Sustainability can be a powerful motivation for change in engineering designs, technologies, processes, and products. This principle emphasizes the importance of being innovative (i.e., "outside-the-box" thinking) in the development of new technologies. The knowledge gained through considering the many dimensions of sustainability should be reflected in how engineering designs accomplish societal objectives.

Principle 9: Actively engage communities and stakeholders in development of engineering solutions. There are many examples of stakeholder and community engagement in the development of engineered solutions in a wide range of activities, including city planning, infrastructure development, and production of manufactured goods. One illustrative example is in the mining industry where *Seven Questions to Sustainability* from the Mining Minerals Sustainable Development North America project (IISD, 2002) has been adopted by key members of this industry. Central to these mining project questions is community engagement from project inception to mine closure. During this engagement, the communities surrounding the proposed mine development express their wishes with regard to managing the economic development and any concerns over local environmental consequences.

A second set of engineering design principles, the 12 Principles of Green Engineering, from Anastas and Zimmerman (2003), is presented here. Each of these principles will be elaborated on in the following pages.

Principle 1: Designers need to strive to ensure that all material and energy inputs and outputs are as inherently nonhazardous as possible. This principle recognizes that significant costs and hazards result from the selection of sources of materials and energy. Additional control systems are required to capture and destroy hazardous materials during production, use, and disposal, all of which add to the cost of the design. If inputs to the system are inherently less hazardous, the risks of failure will be reduced and the amount of resources expended on control, monitoring, and containment will be less.

Principle 2: It is better to prevent waste than to treat or clean up waste after it is formed. The creation of waste in engineered systems adds to the complexity,

---

Principles of Green Engineering

Principle 1: Designers need to strive to ensure that all material and energy inputs and outputs are as inherently nonhazardous as possible.

Principle 2: It is better to prevent waste than to treat or clean up waste after it is formed.

Principle 3: Separation and purification operations should be designed to minimize energy consumption and materials use.

Principle 4: Products, processes, and systems should be designed to maximize mass, energy, space, and time efficiency.

Principle 5: Products, processes, and systems should be "output pulled" rather than "input pushed" through the use of energy and materials.

Principle 6: Embedded entropy and complexity must be viewed as an investment when making design choices on recycle, reuse, or beneficial disposition.

Principle 7: Targeted durability, not immortality, should be a design goal.

Principle 8: Design for unnecessary capacity or capability (e.g., "one size fits all") solutions should be considered a design flaw.

Principle 9: Material diversity in multicomponent products should be minimized to promote disassembly and value retention.

Principle 10: Design of products, processes, and systems must include integration and interconnectivity with available energy and materials flows.

Principle 11: Products, processes, and systems should be designed for performance in a commercial "afterlife."

Principle 12: Material and energy inputs should be renewable rather than depleting.

---

effort, and expense of the design. This is especially true for hazardous wastes, which require extraordinary measures for their control, monitoring, transport, and disposal. To reduce waste generation, the design must strive to incorporate as much of the input materials as possible into final products. This strategy can be applied at many scales, for example, at the molecular level in the design of chemical reactions, at larger scales such as in machining of parts, and further in assembly of discrete parts. Any waste that is generated should be considered as raw material to be used again in the current product system or as input to a separate product system.

Principle 3: Separation and purification operations should be designed to minimize energy consumption and materials use. In the chemical- and mineral-processing industries, large-scale separation processes are among the largest energy-consuming units and generate a significant proportion of emissions and wastes. Even in industry sectors where the mass of the products produced is not large, separation processes can be significant. In electronics manufacturing, the generation of ultra-pure water and the creation of ultra-clean work environments require separation processes with significant costs and energy demands. Design for efficient separation is very important for these industries, and several approaches can be investigated during design.

Gains in energy efficiency can be attempted through heat integration by considering all streams needing to gain or lose energy in the process and also outside the process if in close proximity to other facilities. Similarly, pollution can be prevented by considering mass integration, taking waste streams from one process and using them as raw materials for another. In certain instances products can be induced to self-separate by adjusting conditions to take advantage of physical and chemical properties of the chemicals. As another example, in mechanical systems, reversible fasteners can be used to encourage the disassembly of manufactured parts at the end of life.

Principle 4: Products, processes, and systems should be designed to maximize mass, energy, space, and time efficiency. Energy and mass efficiency were dealt with in Principle 3, so this discussion will focus on space and time efficiency. Space and time are interrelated in many engineered systems, but most obviously in transport of raw material and products. Reducing the distance between points of use for materials can save time and reduce pollution. Close proximity can facilitate exchanges of waste heat and materials in highly integrated production systems. Colocation of manufacturing and recycle facilities can also lead to efficiency gains in many production systems. Industrial parks near residential areas can lead to sharing of excess heat with communities. However, these proximity opportunities that take advantage of space and time factors also must consider safety concerns due to potential exposure to emissions and industrial accidents. The excess time that a product sits in inventory can run up against storage stability limits and could lead to excess waste generation.

Principle 5: Products, processes, and systems should be "output pulled" rather than "input pushed" through the use of energy and materials. It is well known that some chemical reactions can be "pulled" to completion by removing certain co-products from the reaction mixture. This chemical phenomenon, which is termed *Le Châtelier's principle*, can be applied to engineering design across scales of production. "Just-in-time" manufacturing is an example of this principle where only the necessary units are produced in the necessary quantities at the necessary time by bringing production rates exactly in line with demand. Planning manufacturing systems for final output eliminates the wastes associated with overproduction, waiting time, processing, inventory, and resource inputs.

Principle 6: Embedded entropy and complexity must be viewed as an investment when making design choices on recycle, reuse, or beneficial disposition. Entropy and complexity are related concepts when considering engineered systems. Products having a high degree of order and structure are at a low state of entropy. The higher the degree of complexity and structure in a product, the greater is the amount of energy invested to create such structure and complexity. When considering end-of-life options for products, the degree of complexity and structure should point the way to proper reuse, recycle, and remanufacturing options. Highly complex parts should be reused if at all possible in order to avoid the investment required to create a replacement part from virgin (newly extracted from nature) resources.

Principle 7: Targeted durability, not immortality, should be a design goal. Many products last much longer than the expected commercial life. There can be multiple impacts from this extended durability. For example, buildings with inefficient energy systems, designed when energy was relatively inexpensive, may become inoperable in times of expensive energy. At a different scale, the challenge in the design of molecules and of manufactured parts is to create products that are durable yet do not persist indefinitely in the environment. Durability means that the products last for the intended commercial life and are thereafter readily reconfigured or degraded at the end of life into harmless substances that assimilate easily into natural cycles.

Principle 8: Design for unnecessary capacity or capability (e.g., "one size fits all") solutions should be considered a design flaw. Most products are overdesigned to cover a wide application range and settings. Automobiles must be designed to function not only in warm temperatures but also in extremes of cold. However, there are instances where design for "one size fits all" does not make the most sense and is potentially wasteful. For example, a jacket could be designed for the coldest possible climate, but this garment would not be much use for a stroll along the beach on a breezy evening in Florida in the winter. Likewise, lighting of a large classroom, office space, or a room at home with a single light switch would not make as much sense if only one person were in the room at a given time. In such a case, district lighting would save on energy when only one or a few occupants are present in a large space.

Principle 9: Material diversity in multicomponent products should be minimized to promote disassembly and value retention. This design principle has elements in common with Principle 6. Increasing material diversity in products has the effect of making recycling more difficult and expensive because the number of recycling options and their complexity increase as material diversity increases. Different kinds of materials and the use of different additives have a strong influence on recycling methods and costs.

Principle 10: Design of products, processes, and systems must include integration and interconnectivity with available energy and materials flows. This design principle states that products, processes, and entire engineered systems should be designed to use the existing infrastructure of energy and material flows. Integration with existing infrastructure can occur at the scale of a unit operation, production line, manufacturing facility, or industrial park. Taking advantage of existing energy and material flows will minimize the need to generate energy and/or acquire and process raw materials. Applications of this principle include the recovery and use of heat from exothermic chemical reactions, the cogeneration of heat and power, and the recovery of electrical energy by regenerative braking in hybrid vehicles.

Principle 11: Products, processes, and systems should be designed for performance in a commercial "afterlife." Designing components for a second, third, or even longer life is an important strategy in product design. When components are recovered and reused in next-generation products, the environmental impacts of raw material acquisition from virgin resources and conversion are eliminated, and the overall life-cycle environmental impacts are reduced. This strategy is especially important for products that become obsolete prior to component failure, such as cell phones and other electronic devices. Important examples of this principle also include the recovery and recycle of spent copy toner cartridges, the renovation of industrial buildings for housing, and reuse of beverage containers as practiced in Germany, where bottles are more substantial in their construction to allow for collection, washing, sterilization, refilling, and relabeling.

Principle 12: Material and energy inputs should be renewable rather than depleting. The use of nonrenewable raw materials from nature in the design of engineered systems moves the Earth system incrementally toward depletion of finite resources and is therefore unsustainable by definition. All renewable resources derive their usefulness and energy from the sun, and as a result these system inputs can be sustainable, if used at a level consistent with their rate of renewal, for the foreseeable future. Biomass is an important form of renewable resources in that it can serve as not only an energy source but also as a feedstock for design of material products. One important form of a biomass product is liquid transportation fuels. Biofuels are renewable on relatively short timescales, and the cycling of biofuel carbon between the atmosphere, biomass/biofuel, and back to the atmosphere again is readily integrated into natural cycles in a way that might not cause accumulation of CO_2 in the atmosphere (see Chapter 6). An offshoot of this principle is the importance of design for products and systems that integrate well with natural cycles of elements across the life cycle, from raw material acquisition to end-of-life processes.

These engineering design principles, in addition to the others mentioned in the beginning of this section, establish a framework for designing more sustainable products and processes. At first, changes in engineering design are likely to be improvements to inherently unsustainable products and systems, but over time it is hoped that these design principles will move industry and consumers toward inherently sustainable products and production systems. However, there will be tensions in applying these principles. What if making a process inherently safer requires more energy? What if minimizing water use requires more energy? Engineers are accustomed to addressing trade-offs between objectives, but methods for doing so require measures of performance. In most engineering designs the measure is cost. In delivering a specified level of performance of a product, technology, or service, the goal is to minimize cost. This suggests that one mechanism for incorporating objectives related to sustainability into engineering design is to monetize them. Therefore, the next section of this chapter will examine methods for monetizing environmental impacts. Some indicators of environmental and societal performance will be difficult, or arguably impossible, to monetize. In these cases other approaches will be needed for incorporating objectives into engineering design, and some of these approaches are described in the concluding sections of this chapter.

5.3   ECONOMIC PERFORMANCE INDICATORS

Costs associated with poor environmental and societal performance can be very large. Waste disposal fees, permitting costs, and liability costs can all be substantial. Wasted raw material, wasted energy, and reduced manufacturing throughput are also consequences of wastes and emissions. Corporate image and relationships with workers and communities can suffer if performance is substandard. But how can these costs be quantified?

This section will review the tools available for estimating environmental and societal costs and benefits. These include traditional concepts such as the time value of money, present value, payback period, internal rate of return, and other financial evaluation calculations. Nontraditional tools include methods for monetizing environmental costs that are hidden from normal accounting procedures. We will also touch upon less tangible costs and benefits that can still be monetized.

In general, traditional accounting practices have acted as a barrier to implementation of sustainable engineering projects because they hide the costs of poor environmental and societal performance. Many organizations are now giving more consideration to all significant sources of environmental and societal costs. The principle is that if costs are properly accounted for, management practices that foster good economic and societal performance will also foster superior environmental and societal performance.

5.3.1 Definitions

To keep the discussion presented in this chapter clear, it is useful to define a number of terms, as they will be used in this text. Many of these definitions are drawn from an introduction to environmental accounting prepared by the U.S. Environmental Protection Agency (U.S. EPA, 1995).

Internal costs, or private costs, are costs that are borne by a facility. Costs for materials and labor are examples of internal costs. *External costs*, or societal costs, on the other hand, are the costs to society of the facility's activities. The cost associated with a loss of fishable waters due to pollutants discharged by a facility to a stream is an example of an external cost. Often, environmental fees, regulations, and requirements act to internalize what would have otherwise been an external cost, so that a facility that produces waste must pay to reduce its quantity or toxicity or pay a premium for its disposal. This chapter focuses primarily on internal costs.

A typical management accounting system for a manufacturer would include categories for direct materials and labor (costs that are clearly and exclusively associated with a product or service), manufacturing overhead, sales, general and administrative overhead, and research and development. Environmental and societal expenses can be hidden in any or all of these categories but are charged most often as *overhead*. Overhead costs, as opposed to costs of direct materials and labor for production, are often referred to as *indirect costs* and consist of any costs that the accounting system either pools facility-wide and does not allocate among activities or allocates on the basis of a formula. Overhead generally includes indirect materials and labor, capital depreciation, rent, property taxes, insurance, supplies, utilities, and repair and maintenance. It can also include labor costs ranging from supervisor salaries to janitorial services. Often, even the direct environmental costs that could be assigned to a particular process, product, or activity, such as waste disposal, are lumped together facility-wide. This is often done because of practices such as using a single waste disposal company to manage all of a facility's waste. Other environmental costs, such as the costs of filling out forms for reporting waste management practices, are also hidden in the overhead category. Because environmental and societal costs are not traditionally allocated to the activity that is generating wastes, some of the benefits of green engineering projects are masked.

Full-cost accounting is a type of managerial accounting in which as many costs as possible are allocated to products, product lines, processes, services, or activities. Full-cost accounting is pursued because it is useful in determining the profitability of processes and products and in setting prices. Even though full-cost accounting does not focus particularly on environmental and societal costs, it promotes improved performance because the costs of producing waste and societal impacts for individual processes or products are revealed, providing management with a better idea of true costs.

Activity-based costing is similar to full-cost accounting except that the costs are allocated to specific measures of activity. For example, in activity-based costing, the cost of generating a particular kind of waste per pound of production might be measured. Another example would be determining the cost of chemical inputs per item for painting.

These are the primary terms that will be used in this section. It is useful to keep in mind that precise definitions remain in flux and vary from organization to organization, so the terminology used in this chapter is not universal. Nevertheless, it is generally recognized that in environmental and societal cost accounting, words like *full* (e.g., full-cost accounting), *total* (e.g., total cost assessment), *true*, and *life cycle* (e.g., life-cycle costing) are used to indicate that not all costs are captured in traditional accounting and capital budgeting practices.

5.3.2 Estimates of Environmental Costs

The definitions in the previous section made clear that not all environmental costs are captured in traditional accounting and capital budgeting practices. Nevertheless, some measures of environmental and societal costs are available, providing a rough indication of the magnitude of costs and the variation of those costs among industry sectors.

Among the easiest environmental costs to track are the costs associated with treating emissions and disposing of wastes. Direct costs of pollution abatement are tracked by the U.S. Census Bureau and have been increasing steadily. Expenditures in 1972 totaled $52 billion (in 1990 dollars) and were projected to grow to approximately $140 billion (1990 dollars), or 2.0% to 2.2% of gross national product, in the year 2000 (for a review and analysis of these data, see U.S. Congress, Office of Technology Assessment, 1994).

These expenditures are not distributed uniformly among industry sectors. As shown in Table 5-1, sectors such as petroleum refining and chemical manufacturing spend much higher fractions of their net sales and capital expenditures on pollution abatement than do other industrial sectors. Therefore, in these industrial sectors, minimizing costs by preventing wastes and emissions will be far more strategic an issue than in other sectors.

Table 5-1 Pollution Abatement Expenditures by U.S. Manufacturing Industries

Industry Sector	Pollution Control Expenditures (as % of Sales)	Pollution Control Expenditures (as % of Value Added)	Pollution Control Capital Expenditures (as % of Total Capital Expenditures)
Petroleum	2.25%	15.42%	25.7%
Primary metals	1.68%	4.79%	11.6%
Paper	1.87%	4.13%	13.8%
(pulp mills)	(5.70%)	(12.39%)	(17.2%)
Chemical manufacturing	1.88%	3.54%	13.4%
Stone products	0.93%	1.77%	7.2%
Lumber	0.63%	1.67%	11.1%
Leather products	0.65%	1.37%	16.2%
Fabricated materials	0.65%	1.34%	4.6%
Food	0.42%	1.11%	5.3%
Rubber	0.49%	0.98%	2.0%
Textile	0.38%	0.93%	3.3%
Electric products	0.49%	0.91%	2.9%
Transportation	0.33%	0.80%	3.0%
Furniture	0.38%	0.73%	3.4%
Machinery	0.25%	0.57%	1.9%

Source: Data reported by U.S. Congress, Office of Technology Assessment, 1994; original data collected by U.S. Census Bureau

---

Example 5-1   Potential pollution control costs for greenhouse gases

Estimate control costs for greenhouse gases for an electricity-generating unit (EGU or power plant) as a percentage of sales revenues. Assume that (1) the EGU uses coal as a fuel and converts the heat of combustion of coal into electricity with 35% efficiency, (2) the heating value of coal is 10,000 BTU/lb, (3) coal is 85% carbon, (4) carbon costs $20 per ton to capture and sequester, and (5) electricity can be sold for $0.10 per kilowatt-hour.

Solution: Electricity generated per ton of carbon:

1 ton coal/0.85 ton C * 2000 lb/ton * 10,000 BTU/lb coal * 1 kWh/3412 BTU * 0.35 = 2415 kWh/ton C
The control costs are $20 for 2415 kWhr of generation.
This much generation leads to $241 in electricity sales, so the control costs are about 8% of sales.

Pollution abatement costs reported by individual companies both reflect these general trends and provide more detail about the magnitude and the distribution of environmental expenditures. For example, Tables 5-2 and 5-3 show the distribution of environmental costs reported by the Amoco Yorktown refinery and DuPont's LaPorte chemical manufacturing facility (Heller et al., 1995; Shields et al., 1995). In the case of the Amoco refinery, only about a quarter of the quantified environmental costs are associated with waste treatment and disposal; the costs are summarized in Table 5-1. Costs associated with removing sulfur from fuels, meeting other environmentally based fuel requirements, and maintaining environmental equipment were greater than the costs associated with waste treatment and disposal. This indicates that the magnitude of environmental costs is substantially greater than that reported in Table 5-1, and that these costs may be hard to identify.

Table 5-2 Summary of Environmental Costs at the Amoco Yorktown Refinery

Cost Category	Percentage of Annual Non-Crude Operating Costs
Waste treatment	4.9%
Maintenance	3.3%
Product requirements	2.7%
Depreciation	2.5%
Administration, compliance	2.4%
Sulfur recovery	1.1%
Waste disposal	0.7%
Fees, fines, penalties	0.2%
Total costs	21.9%

Source: Heller et al., 1995

Table 5-3 Summary of Environmental Costs at the DuPont LaPorte Chemical Manufacturing Facility

Cost Category	Percentage of Manufacturing Costs
Taxes, fees, training, legal	4.0%
Depreciation	3.2%
Operations	2.6%
Contract waste disposal	2.4%
Utilities	2.3%
Salaries	1.8%
Maintenance	1.6%
Engineering services	1.1%
Total	19.1%

Source: Shields et al., 1995

Table 5-3 shows that the profile of environmental costs at a DuPont chemical manufacturing facility exhibits many of the same characteristics. Waste treatment and disposal costs are less than a quarter of the annual, quantifiable, environmental costs.

Taken together, Tables 5-1 to 5-3 demonstrate that environmental costs are substantial but that quantifying these costs will be challenging. The next several sections of this chapter will present a framework and procedures for estimating these costs.

5.3.3   A Framework for Evaluating Environmental Costs

Engineering projects are generally not undertaken unless they are financially justifiable, and projects designed to improve environmental and societal performance beyond regulatory requirements usually must compete financially with all other projects under consideration at a facility. Fortunately, projects resulting in improved environmental and societal performance are frequently profitable. However, the potential profitability of environmental and societal projects is difficult to assess, and it is common for many of the financial benefits of improved environmental and societal performance to be neglected when projects are analyzed. That is why a better understanding of methods for estimating these costs and benefits serves to promote sustainable engineering.

In this section, the types and magnitudes of costs associated with emissions, waste generation, and societal impacts are described and categorized. Five categories, or tiers, of costs will be considered, following the framework recommended in the Total Cost Assessment Methodology developed by the American Institute of Chemical Engineers' Center for Waste Reduction Technologies (AIChE CWRT, 2000). These are

- **Tier I:** Costs normally captured by engineering economic evaluations
- **Tier II:** Administrative and regulatory environmental costs not normally assigned to individual projects
- **Tier III:** Liability costs
- **Tier IV:** Costs and benefits, internal to a company, associated with improved environmental and societal performance
- **Tier V:** Costs and benefits, external to a company, associated with improved environmental and societal performance

Tier I costs are the types of costs quantified in traditional economic analyses. Here are some specific examples:

Costs Traditionally Evaluated in Financial Analyses

- Capital equipment
- Materials
- Labor

- Supplies
- Utilities
- Structures
- Salvage value

As discussed, traditional accounting systems that focus on Tier I costs fail to capture some types of environmental costs. Here are examples of some of the costs that are frequently overlooked by traditional methods:

Environmental Costs Often Charged to Overhead

- Off-site waste management charges
- Waste treatment equipment
- Waste treatment operating expenses
- Filing for permits
- Taking samples
- Filling out sample reporting forms
- Conducting waste and emission inventories
- Filling out hazardous waste manifests
- Inspecting hazardous waste storage areas and keeping logs
- Making and updating emergency response plans
- Sampling storm water
- Making chemical usage reports (some states)
- Reporting on pollution prevention plans and activities (some states)

These costs are generally charged to overhead and therefore may be "hidden" when project costs are evaluated. These will be referred to as Tier II or hidden costs. Note that these costs are actually borne by facilities regardless of whether facilities choose to quantify them or assign them to project or product lines.

A less tangible set of costs are those designated as Tier III—liability costs. An accounting definition of *liability* is a "probable future sacrifice of economic benefits arising from present obligations to transfer assets or provide services in the future" (Financial Accounting Standards, 1985; Institute of Management, 1990). Liability costs could include

- Compliance obligations
- Remediation obligations
- Fines and penalties
- Obligations to compensate private parties for personal injury, property damage, and economic loss
- Punitive damages
- Natural resource damages

A final set of costs are designated as Tier IV or Tier V, which can be referred to as image, relationship, or societal costs (AIChE CWRT, 2000). These costs arise in relationships with customers, investors, insurers, suppliers, lenders, employees, regulators, and communities. They are perhaps the most difficult to quantify.

Thus, a basic framework for estimating costs and benefits associated with environmental activities consists of five tiers, beginning with the most tangible costs and extending to the least quantifiable costs. A more thorough treatment of environmental cost accounting is given in Allen and Shonnard (2012), and the interested reader is referred to that text for calculation methods.

5.4   ENVIRONMENTAL PERFORMANCE INDICATORS

Sustainable engineering design principles were discussed in Section 5.2, and these principles stand as guidelines for engineers to create more sustainable processes and products. But in order to fully use these principles in design, indicators of environmental performance are needed to quantify the impacts and compare impacts to conventional technologies, processes, and products. This approach to environmental management is based on the principle that "anything that can be measured can be improved." The purpose of this section is to introduce environmental performance measures that have been used in engineering design and product development.

Chapter 2 introduced concepts of risk assessment and included applications to human health and to product life-cycle assessment. In both of these applications, a methodology was used that first defined a system to evaluate, estimated environmental releases, determined exposure to sensitive human and environmental receptors, and calculated damage to human health or impacts to the environment. We will further develop those approaches in this section.

In proceeding further it is most convenient to discuss environmental indicators in the context of a product life-cycle assessment. The ideas to be presented can also be applied to the evaluation of a process or any human activity that takes resources from nature and emits by-products back to nature. Consider Figure 5-2, which describes a generic product system in which raw materials are converted to chemical intermediates and then to a final product, which is then used and disposed of or reused. An inventory of material and energy inputs is developed at each stage, and an accounting of output wastes is also included. Wastes that are not captured in pollution control devices reach the environment, causing a number of midpoint damage effects to the environment itself, such as warming the climate or destroying ozone in the stratosphere. A series of other environmental processes are involved in transforming midpoint effects to endpoint damage to humans and ecosystems. These endpoint damages can cause shortening of human life expectancy or loss of plant or animal species from ecosystems.

The differences between midpoint and endpoint indicators of environmental impact are further illustrated in Figure 5-3, which shows the environmental mechanisms starting with emissions, moving to midpoint effects, and ultimately to a number

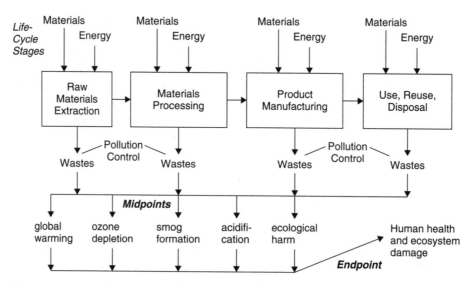

Figure 5-2 A generic product life cycle showing key stages in a product system, key inputs and wastes, effects of emitted wastes on midpoint damages in the environment, and endpoint damage to humans and ecosystems

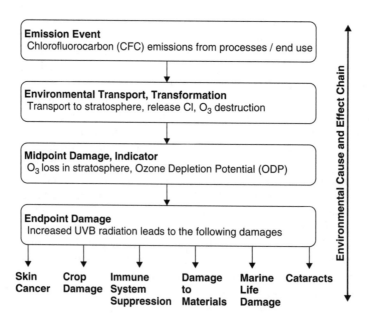

Figure 5-3 Environmental mechanisms involved in conversion of emissions of CFCs and halons into midpoint and endpoint damages (Adapted from Bare et al., 2003)

of endpoint damages. Both midpoint and endpoint impact indicators have advantages and limitations. Midpoint indicators are closer to emissions in the environmental cause-and-effect chain, are easier to model, have less uncertainty, but are less relevant to decision makers who may not understand the true endpoint effects. Endpoint indicators are farther down the environmental cause-and-effect chain, are more difficult to model, have a higher degree of uncertainty, but are more relevant to decision makers. It is often desirable to present both types of indicators when evaluating engineering designs (Bare et al., 2000). In this section, we will focus on midpoint indicators of environmental impacts and refer the interested reader to other publications describing endpoint damage methods (Bare et al., 2000; Goedkoop and Spriensma, 1999).

5.4.1   Life-Cycle Impact Assessment

The emissions and wastes from a product life cycle constitute the "inventory" upon which environmental impact indicators are based. Beyond the inventory of emissions, lists of raw materials and primary energy sources are added, as well as other categories such as land use, but these categories beyond emissions will be set aside for this section. Life-cycle inventories do not by themselves characterize the environmental performance of a product, process, or service. This is because overall quantities of wastes and emissions as well as raw material and energy requirements must be considered in conjunction with their potency of effect on the environment. Simply stated, a pound of lead emitted to the atmosphere has a different environmental impact from a pound of iron emitted to surface waters. To develop an overall characterization of the environmental performance of a product or process, throughout its life cycle, requires that life-cycle inventory data be converted into estimates of environmental impact.

The process of producing life-cycle impact assessments is generally divided into three major steps (Fava et al., 1992). They are

- **Classification,** where inputs and outputs determined during the inventory process are classified into environmental impact categories; for example, methane, carbon dioxide, and CFCs would be classified as global warming gases.
- **Characterization,** where the potency of effect of the inputs and outputs on their environmental impact categories is determined; for example, the relative global warming potentials (Table A5-1) of methane, carbon dioxide, and CFCs would be identified in this step.
- **Valuation,** where the relative importance of each environmental impact category is assessed, so that a single index for environmental performance can be calculated. This subject is covered in other texts (see Allen and Shonnard, 2012) and will not be covered here.

Note that the classification and characterization steps are generally based on scientific data or models. The data may be incomplete or uncertain, but the process of classification and characterization is generally objective. In contrast, the valuation step is inherently subjective and depends on the value that decision makers in society place on various environmental impact categories.

Each of the three steps is discussed in more detail in the following sections.

Classification

As a first step in life-cycle impact assessment, inputs and outputs that were the subject of the inventory are classified into environmental impact categories, such as the following:

- Global warming
- Stratospheric ozone depletion
- Photochemical smog formation
- Human carcinogenicity
- Atmospheric acidification
- Aquatic toxicity
- Terrestrial toxicity
- Habitat destruction
- Depletion of nonrenewable resources
- Eutrophication

Note that some impact categories might apply to very local phenomena (for example, aquatic toxicity to organisms found only in certain ecosystems), while other impact categories are global (for example, stratospheric ozone depletion and global warming).

As an example of classification, consider the list of air emissions inventoried for a study that compared glass and polyethylene, which is given in Table 5-4. Carbon monoxide emissions are higher for polyethylene than for glass on a mass basis, while emissions of nitrogen oxides are higher for glass than for polyethylene. Nitrogen oxide emissions would be classified as photochemical smog precursors, global warming gases, and acid precipitation and acid deposition precursors. Carbon monoxide emissions, on the other hand, would be classified as a smog precursor.

Table 5-4 Selected Air Emissions from the Production of 1 kg of Polyethylene and 1 kg of Glass

	kg Emissions per kg of Polyethylene	kg Emissions per kg of Glass
Nitrogen oxides	0.0011	16
Sulfur dioxide	0.00099	0.0027
Carbon monoxide	0.00067	0.000057

Source: Adapted from Allen et al., 1992

Characterization

The second step of impact assessment, characterization, generally consists of assigning relative weights or potencies to different types of emissions, energy use, and materials use. These potencies reflect the degree to which the inventory elements contribute to environmental impacts. For example, if the impact category is global warming, then relative global warming potentials can be used to weight the relative impact of emissions of different global warming gases. Once these *potency factors* are established, the inventory values for inputs and outputs are combined with the potency factors to arrive at *impact scores*.

Although no single methodology has gained universal acceptance, several useful methodologies for indexing environmental and health impacts of chemicals have recently appeared in the literature. Many of the indexing methods include metrics for abiotic as well as biotic impacts. In the abiotic category, global warming, stratospheric ozone depletion, acidification, eutrophication, and smog formation are often included. In the biotic category, human health and plant, animal, and other organism health are impacts of concern. For issues of environmental and economic sustainability, resource depletion indexes reflect long-term needs for raw materials utilization. A review of several of these methodologies would indicate that many environmental indicators (indexes) have been constructed by employing separate parameters for the inherent impact potential (IIP) and exposure potential (EP) of an emitted chemical. The index is normally expressed as a product of inherent impact and exposure, following risk assessment guidelines (NRC, 1983; Heijungs et al., 1992; SETAC, 1993).

The general form of a dimensionless environmental risk index (indicator) is defined as

$$(\text{Dimensionless Risk Index})_i = \frac{[(EP)(IIP)]_i}{[(EP)(IIP)]_B} \tag{5-1}$$

where B stands for the benchmark compound and i the chemical of interest. To estimate the indicator I for a particular impact category due to all of the chemicals released from a product system, we must sum the contributions for each chemical weighed by their emission rate:

$$I = \sum_i (\text{Dimensionless Risk Index})_i \times m_i \tag{5-2}$$

The following is a brief summary of environmental and health indexes that have been used to compare impacts of chemicals, processes, or products.

Global Warming. A common index for global warming is the global warming potential (GWP), which is the time-integrated climate forcing from the release of 1 kg of a greenhouse gas relative to that from 1 kg of carbon dioxide (IPCC, 2007):

$$GWP_i = \frac{\int_0^{TH} a_i c_i \, dt}{\int_0^{TH} a_{co_2} c_{co_2} \, dt} \tag{5-3}$$

where a_i is the predicted radiative forcing of gas i (Wm^{-2}) (which is a function of the chemical's infrared absorbance properties and C_i), C_i is its predicted concentration in the atmosphere (parts per billion), and TH is the number of years over which the integration is performed (time horizon), which is chosen to be 20, 100, or 500 years to model short-term or long-term warming potentials. The concentration is a function of time (t), primarily due to loss by chemical reaction. GWP contains an exposure factor (C_i) and an impact factor (a_i). Several authors have developed models to calculate GWP, and as a result, variations in GWP predictions have appeared (Fisher et al., 1990a; Derwent, 1990; Lashof and Ahuja, 1990; Rotmans, 1990). A list of "best estimates" for GWPs has been assembled from these model predictions by a panel of experts convened under the Intergovernmental Panel on Climate Change (IPCC, 2007). Table A5-1 is a list of GWPs for several important greenhouse gases. The GWP for each chemical is influenced mostly by the chemical's tropospheric residence time and the strength of its infrared radiation absorbance (band intensities). All of these gases are extremely volatile, do not dissolve in water, and do not adsorb to soils and sediments. Therefore, they will persist in the atmosphere after being released from sources.

The product of the GWP and the mass emission rate of a greenhouse gas results in the equivalent emission of carbon dioxide, the benchmark compound, that would result in the same radiative forcing. By performing this calculation, a direct connection is made between mass emission from a process of any greenhouse gas and global warming impact. The global warming index for the entire product system is the sum of the emissions-weighted GWPs for each chemical:

$$I_{GW} = \sum_i \left(GWP_i \times m_i\right) \tag{5-4}$$

where m_i is the mass emission rate of chemical i from the entire product system (kg/hr). This step will provide the equivalent process emissions of greenhouse chemicals in the form of the benchmark compound, CO_2.

The global warming index as calculated in equation (5-4) accounts for direct effects of the chemical, but most chemicals of interest are so short-lived in the atmosphere (because of the action of hydroxyl radicals in the troposphere) that they disappear (become converted to CO_2) long before any significant direct effect can be felt. However, organic chemicals will have an indirect global warming effect because of the carbon dioxide released upon oxidation within the atmosphere and other compartments of the environment. In order to account for this indirect effect for organic compounds with atmospheric reaction residence times of *less than half a year*, an indirect GWP is defined as (Shonnard and Hiew, 2000)

$$GWP(\text{indirect}) = N_c \frac{MW_{co_2}}{MW_i} \tag{5-5}$$

where N_C is the number of carbon atoms in the chemical i and the molecular weights (MW) convert from a molar to a mass basis for GWP, as originally defined. Organic chemicals whose origins are in renewable biomass (plant materials) will have no net global warming impact because the CO_2 released upon environmental oxidation of these compounds will replace CO_2 removed from the atmosphere during photosynthesis of the biomass (see Chapter 6). Example 5-2 demonstrates the application of equation (5-4) to the production of an important industrial solvent.

Example 5-2   Global warming index for air emissions of 1,1,1-trichloroethane from a production process

1,1,1-Trichloroethane (1,1,1-TCA) is used as an industrial solvent for metal cleaning, as a reaction intermediate, and for other important uses (WHO, 2000). Sources for air emissions include distillation condenser vents, storage tanks, handling and transfer operations, fugitive sources, and secondary emissions from wastewater treatment. This example will estimate the global warming impact of the air emissions from this process. Include direct impacts to the environment (from 1,1,1-TCA) and indirect impacts from energy usage (CO_2 and N_2O release) in your analysis. The following data show the major chemicals that impact global warming when emitted from the process.

Determine the global warming index for the process and the percentage contribution for each chemical.

Data: Air Emissions (Based on a 15,500 kg 1,1,1-TCA/hr Process)

Chemical	m_i (kg/hr)	GWP_i
TCA	10	100
CO_2	7,760	1
N_2O	0.14	298

Source: U.S. EPA, 1979–1991; Allen and Rosselot, 1997; Boustead, 1993

Solution: Using equation (5-4), the process global warming index is

$$I_{GW} = (10 \text{ kg/hr})(100) + (7760 \text{ kg/hr})(1) + (0.14 \text{ kg/hr})(298)$$
$$= 1000 + 7760 + 41.7$$
$$= 8801.7 \text{ kg/hr}$$

The percent of the process I_{GW} for each chemical is

$$1,1,1\text{-TCA: } (1000/8801.7) \times 100 = 11.4\%$$
$$CO_2 : (7760/8801.7) \times 100 = 88.1\%$$
$$N_2O: (43.4/8801.7) \times 100 = 0.5\%$$

Discussion   In this case study, the majority of the global warming impact from the production of 1,1,1-TCA is from the energy requirement of the process and not from the emission of the chemical with the highest global warming potential. This analysis assumes that a fossil fuel was used to satisfy the energy requirements of the process. If renewable resources were used (biomass-based fuels), the impact of CO_2 on global warming would be significantly reduced. Finally, the majority of the global warming impact of 1,1,1-TCA could very well be felt during the use stage of its life cycle if the compound evaporates when it is used. This possibility was not included in this example.

Ozone Depletion. The ozone depletion potential (ODP) of a chemical is the predicted time- and height-integrated change ($\delta[O_3]$) in stratospheric ozone caused by the release of a specific quantity of the chemical relative to that caused by the same quantity of a benchmark compound, trichlorofluoromethane (CFC-11, CCl_3F) (Fisher et al., 1990b):

$$ODP_i = \frac{\delta[O_3]_i}{\delta[O_3]_{CFC-11}} \tag{5-6}$$

Model calculations for ODP have been carried out using one- and two-dimensional photochemical models. A list of ODPs, for a small number of chemicals, has been assembled by a committee of experts (WMO, 2007; U.S. EPA, 2011). The product of the ODP and the mass emission rate of a chemical i results in the equivalent impact of an emission of CFC-11. Tables A5-2 and A5-3 show a list of ozone depletion potential values for important industrial compounds. Notice that the brominated compounds in these tables have much larger ODPs than the chorinated species. Also, it is thought that fluorine does not contribute to ozone depletion (Ravishankara et al., 1994). Like the global warming chemicals in Table A5-1, the chemicals in Tables A5-2 and A5-3 will exist almost exclusively in the atmosphere after being emitted by sources. The ozone depletion index for an entire product system is the sum of all contributions from emitted chemicals multiplied by their emission rates. The equivalent emission of CFC-11 for the entire product system is then

$$I_{OD} = \sum_i (ODP_i \times m_i) \tag{5-7}$$

Acid Rain. The potential for acidification for any compound is related to the number of moles of H^+ created per number of moles of the compound emitted. The balanced chemical equation can provide this relationship:

$$X + \bullet\bullet\bullet\bullet\bullet \rightarrow \alpha H^+ + \bullet\bullet\bullet\bullet\bullet \tag{5-8}$$

where X is the emitted chemical substance that initiates acidification and a (moles H^+/mole X) is a molar stoichiometric coefficient. Acidification is normally expressed on a mass basis, and therefore the H^+ created per mass of substance emitted (η_i, moles H^+/kg i) is

$$\eta_i = \frac{\alpha_i}{MW_i} \tag{5-9}$$

where MW_i is the molecular weight of the emitted substance (kg i/moles i). As before, we can introduce a benchmark compound (SO_2) and express the acid rain potential (ARP_i) of any emitted acid-forming chemical relative to it (Heijungs et al., 1992):

$$ARP_i = \frac{\eta_i}{\eta_{so_2}} \qquad (5\text{-}10)$$

The number of acidifying compounds emitted by industrial sources is limited to a rather small number of combustion by-products and other precursor or acidic species emitted directly to the environment. Table A5-4 lists the acid rain potentials for several common industrial pollutants. The total acidification potential index of an entire product system is defined similarly to I_{GW} and I_{OD}:

$$I_{AR} = \sum_i (ARP_i \times m_i) \qquad (5\text{-}11)$$

Smog Formation. A scientifically based smog formation assessment tool will aid in identifying opportunities for reducing smog formation. The most important process for ozone formation in the lower atmosphere is photodissociation of NO_2:

$$NO_2 + h\nu \longrightarrow O(^3P) + NO$$
$$O(^3P) + O_2 + M \longrightarrow O_3 + M$$
$$O_3 + NO \longrightarrow NO_2 + O_2$$

where M is nitrogen or molecular oxygen. This cycle results in O_3 concentration being in a photostationary state dictated by the NO_2 photolysis rate and ratio of $[NO_2]/[NO]$. The role of VOCs is to form radicals that can convert NO to NO_2 without causing O_3 destruction, thereby increasing the ratio $[NO_2]/[NO]$ and increasing O_3.

$$VOC + {}^\bullet OH \longrightarrow {}^\bullet RO_2 + \text{other oxidation products}$$
$${}^\bullet RO_2 + NO \longrightarrow NO_2 + \text{radicals}$$
$$\text{radicals} \longrightarrow {}^\bullet OH + \text{other oxidation products}$$

The tendency of individual VOCs to influence O_3 levels depends upon their hydroxyl radical (${}^\bullet OH$) rate constant and elements of their reaction mechanism, including radical initiation, radical termination, and reactions that remove NOx. Simplified smog formation potential indexes have been proposed based only on VOC hydroxyl radical rate constants, but these have not correlated well with model predictions of photochemical smog formation (Allen et al., 1992; Japar et al., 1991).

Incremental reactivity (IR) has been proposed as a method for evaluating smog formation potential for individual organic compounds. It is defined as the change in grams of ozone formed as a result of emission into an air shed of 1 g of the VOC (Carter and Atkinson, 1989). Several computer models have been developed to evaluate incremental reactivity (Carter and Atkinson, 1989; Carter, 1994; Chang and Rudy, 1990; Dodge, 1984). In general, predicted VOC incremental reactivities are greatest when NOx levels are high relative to reactive organic gases (ROG) and lowest (or even negative) when NOx is relatively low. Therefore, the ratio ROG/NOx is an important model parameter. Lists of incremental smog formation reactivities for many VOCs have been compiled (Carter, 1994; Heijungs et al., 1992). An estimation methodology has also been developed which circumvents the need for computer model predictions, though the practical use of this method is limited because of lack of detailed smog reaction mechanisms for a large number of compounds (Carter and Atkinson, 1989). Although several reactivity scales are possible, the most relevant for comparing VOCs is the maximum incremental reactivity (MIR) which occurs under high NOx conditions when the highest ozone formation occurs (Carter, 1994).

The smog formation potential (SFP) is based on the maximum incremental reactivity scale of Carter (Carter, 1994):

$$SFP_i = \frac{MIR_i}{MIR_{\text{ROG}}} \tag{5-12}$$

where MIR_{ROG} is the average value for background "reactive organic gases," the benchmark compound for this index. This normalized and dimensionless index is similar to the one proposed by the Netherlands Agency for the Environment (Heijungs et al., 1992). Table A5-5 contains a listing of calculated MIR values for many common volatile organic compounds found in fuels, paints, and solvents. Most of the chemicals in Table A5-5 are volatile and will maintain a presence in the atmosphere after release into the air, with the exception of the higher-molecular-weight organics. The total smog formation potential is the sum of the MIRs and emission rates for each smog-forming chemical in the process. The process equivalent emission of ROG is

$$I_{SF} = \sum_i (SFP_i \times m_i) \tag{5-13}$$

Example 5-3 is an application of these environmental impact equations to the generation of electricity using different primary energy sources: coal, oil, natural gas, nuclear, hydro, wind, and photovoltaics.

Example 5-3   Comparing electricity generation by fuel type

Electricity is one of the most important forms of energy that is used in industry, commerce, and private residences. In this example, a list of pollutant emissions to air will be listed for different electricity fuel types. The emissions will be classified into different impact categories and their impact potentials characterized. Finally, environmental indicators will be calculated from the emission and impact potential data.

Solution: Emissions of gaseous pollutants based on the production of 1 kWh of each electricity type are shown in the table below. These inventory data represent a compilation over the entire life cycle from extraction of raw materials from nature to the output of electricity from the power plant: a "cradle-to-gate" inventory.

		Electricity (1 kWh)						
Inventory	Unit	Hard Coal	Oil	Natural Gas	Nuclear	Hydro	Wind	PV[a]
Dinitrogen mon- oxide	mg	29.01	43.06	11.88	0.54	0.06	0.42	1.68
NMVOC[b]	mg	126.14	270.75	184.21	7.44	2.26	5.50	50.32
Methane	g	1.47	0.48	3.41	0.02	0.05	0.03	0.11
Nitrogen oxides	g	2.55	2.75	0.37	0.04	0.01	0.02	0.10
Sulfur dioxide	g	5.27	6.60	5.79	0.05	0.01	0.03	0.15
Carbon dioxide	g	1143.99	860.03	595.07	11.32	3.79	10.40	41.10

[a] PV = photovoltaic
[b] NMVOC = non-methane volatile organic compounds, unspecified

$$\text{Global warming index:} \quad I_{GW} = \sum_i (GWP_i \times m_i)$$

The inventory values for each pollutant are multiplied by the appropriate GWP_i. For example, the table entry (below) for hard coal dinitrogen monoxide is calculated as $(298 \times 29.01 \text{ mg})/(1000 \text{ mg/g}) = 113.52$ g. From this table we see that the global warming index for these electricity types differs by over a factor of 100 from the highest (coal) to the lowest (hydro). CO_2 is the most important greenhouse gas, but N_2O can constitute over 60% of I_{GW}. Oil power achieves about a 25% reduction in global warming impact compared to coal, and natural gas shows nearly 38% in savings.

		Electricity (1 kWh)						
$(GWP_i \times m_i)$	Unit	Hard Coal	Oil	Natural Gas	Nuclear	Hydro	Wind	PV
Dinitrogen monoxide (GWP = 298)	g	113.52	91.11	132.15	7.10	5.01	18.59	37.53
Methane (GWP = 25)	g	36.86	12.00	85.26	0.52	1.29	0.76	2.74
Carbon dioxide (GWP = 1)	g	1143.99	860.03	595.07	11.32	3.79	10.40	41.10
I_{GW}	g	**1294.37**	**963.15**	**812.48**	**18.94**	**10.09**	**29.75**	**81.37**

$$\text{Acidification index:} \quad I_{AR} = \sum_i (ARP_i \times m_i)$$

Similar to the preceding case, the inventory values for each pollutant are multiplied by the appropriate ARP_i. For example, the table entry for hard coal nitrogen oxides (NO) is calculated as $(1.07 \times 2.55 \text{ g}) = 2.72$ g. From this table we see that the acidification index for these electricity types differs by over a factor of 100 from the highest (oil) to the lowest (hydro). SO_2 is the most important compound, but NO can constitute up to 50% of I_{AR}.

| (ARP$_i$ × m$_i$) | Unit | Electricity (1 kWh) | | | | | | |
		Hard Coal	Oil	Natural Gas	Nuclear	Hydro	Wind	PV
Nitrogen oxides (ARP = 1.07)	g	2.72	2.95	0.40	0.05	0.01	0.03	0.11
Sulfur dioxide (ARP = 1.0)	g	5.27	6.60	5.79	0.05	0.01	0.03	0.15
I_{AR}	g	**7.99**	**9.54**	**6.19**	**0.10**	**0.02**	**0.06**	**0.25**

$$\text{Smog formation index:} \quad I_{SF} = \sum_i (SFP_i \times m_i)$$

The inventory values for each pollutant are multiplied by the appropriate SFP_i. For example, the table entry for hard coal NMVOC is calculated as $(3.10 \times 126.14$ mg)/1000 = 0.39 g. From this table we see that the smog formation index for these electricity types differs by nearly a factor of 100 from the highest (oil) to the lowest (hydro). NMVOC is the most important even though emissions of methane are higher.

| (SFP$_i$ × m$_i$) | Unit | Electricity (1 kWh) | | | | | | |
		Hard Coal	Oil	Natural Gas	Nuclear	Hydro	Wind	PV
NMVOC (SFP = 3.10)	g	0.39	0.84	0.57	0.02	0.01	0.02	0.16
Methane (SFP = 0.015)	g	0.02	0.01	0.05	0.00	0.00	0.00	0.00
I_{SF}	g	**0.41**	**0.85**	**0.62**	**0.02**	**0.01**	**0.02**	**0.16**

This comparison shows that electricity generated from coal and oil has the highest impact on the environment, and that renewable power from hydro, wind, and solar (PV) are among the best alternatives based on these indicators. Nuclear also exhibits very low impacts for these indicators, but the same conclusion may not apply if human health indicators and other types of impacts are included. These results can be used to inform decisions about investments in future power production and also in consumer decision making on whether or not to purchase green power.

Human and Ecosystem Health. Environmental indicators such GWP, ODP, ARP, and SFP are based on cause-and-effect mechanisms involving physical and chemical transformation of pollutants. Indicators for human and ecosystem health rely on both objective (physical and chemical) and subjective factors. The TRACI method is one of the more objective, science-based indicator sets for human and ecosystem health (Bare et al., 2003), but similar methods are also available (Guinée et al., 1996; Hertwich et al., 2001). The TRACI method is based on a multimedia fate, multipathway human exposure and toxicological potency approach. Twenty-three exposure pathways were taken into account, including inhalation, ingestion of water and food, and dermal contact with the soil and water. Toxicity is based on cancer potencies for carcinogens and reference doses or concentrations for noncarcinogens (see Chapter 2 for sources of toxicological data). The TRACI method is applicable to long-term and large-scale assessments of health effects. Methods that include toxicological properties but not environmental fate and human exposure modeling are often used, yet they are less scientifically rigorous. One such method classifies toxic releases into six human health categories depending on exposure route (inhalation, ingestion, dermal) and severity of toxic response according to EU Directive 67/546/EEC (Saling et al., 2002; Landsiedel and Saling, 2002). Other methods rely on environmental and occupational health regulatory concentration limits to weight the relative toxicity potentials of compounds (Postlethwaite and de Oude, 1996).

At times, different life-cycle impact systems will lead to different results. As an example, consider an inventory of the releases of organochlorine compounds to the Great Lakes Basin, which was performed by Rosselot and Allen (1999). Three different potency factor schemes for human and ecological toxicity impact categories were used to rank the inventory data. The results of the rankings are shown in Table 5-5. Ideally, each of the potency factor schemes would result in the same rank ordering of chemicals; however, the data show that the different potency schemes lead to different rank ordering of some of the compounds. The potency schemes agree in their rankings of trichloroethylene, 1,2-dichloroethane, and PCBs but disagree in their rankings of dichloromethane, endosulfan, and hexachlorobutadiene. Note that not all of the characterization systems listed in the table were created for the purpose of conducting life-cycle impact assessments. Instead, some of them were developed in order to rank emissions. Also note that ranking these compounds by mass of release (the order in which they are listed in the table) would give very different results from ranking them by potency of effect for any of the characterization schemes. Thus, while not all potency factors lead to identical results, ignoring the concept of potency and considering only the mass of emissions may place too great an emphasis on relatively benign compounds that are emitted in large amounts.

Why would different potency schemes lead to different results? The answer is simple. The methods are often based on different criteria. Some commonly used potency factors (Swiss Federal Ministry for the Environment [BUWAL]; Postlethwaite and de Oude, 1996) are based on data from environmental regulations. In these systems, each emission is characterized based on the volume of air or water that would be required to dilute the emission to its legally acceptable limit. For example, if air quality regulations allowed 1 ppb by volume of a compound in ambient air, then 1 billion moles of air (22.4 billion liters of air at standard temperature and pressure) would be required to dilute 1 mole of the compound to the allowable standard. This volume per unit mass or mole of emission is called the *critical dilution volume* and can vary across political boundaries.

Table 5-5 Rankings of 1993 Releases of Chlorinated Organic Compounds in the Great Lakes Basin for Several Potency Factor Schemes

Compound[a]	EPA Huma Risk[b]	Dutch Human Tox[c]	Dutch Aquatic Tox[c]	EPA Eco Risk[b]	MPCA Tox Score[d]	Dutch Terr Tox[c]
Tetrachloroehtylene	8	4	5	10	6	3
Dichloromethane	13	5	8	10	2	9
Trichloroehtylene	8	12	12	10	10	10
1,1,1-Trichloroethane	8	1	6	10	7	4
Chloroform	6	8	4	6	1	6
1,2-Dichloroethane	8	6	7	8	4	7
PCBs	1	2	2	1	5	1
Endosulfan	2	7	1	2	11	2
Carbon tetrachloride	3	10	10	2	3	11
Vinyl chloride	6	9	13	6	9	13
Chlorobenzene	8	14	14	8	13	14
Benzyl chloride	13	15	15	10	14	15
Hexachlorobutadiene	3	13	11	4	12	8
2,4-Dichlorophenol	13	11	9	10	15	12
2,3,7,8-TCDD	3	3	3	4	8	5

[a] Listed in descending order of quantity released.

[b] U.S. EPA, Waste Minimization Prioritization Tool, used to rank pollutants.

[c] Dutch: Guinée et al. (1996) considers environmental fate and transport, developed specifically for life-cycle assessment.

[d] MPCA: Pratt et al. (1993), Minnesota Pollution Control Agency system for ranking air pollutants. May be based on human or animal effects.

5.5 SOCIAL PERFORMANCE INDICATORS

Engineers often incorporate social concerns in the design of products, infrastructure, and systems by considering the end user functions and interactions. But social impacts within the context of sustainability move well beyond end user functions to the fulfillment of human needs and improvement of the human condition. These needs and social conditions span a range of issues from fulfillment of basic nutritional, sanitation, and security requirements to higher-level human aspirations in areas of health, education, and the arts. But how can engineers incorporate elements of social sustainability into design in a formal fashion? This section will attempt to provide some preliminary answers to this question.

A basic issue to address in developing social indicators of sustainability is what impacts should be considered. There is no general consensus among sustainability experts, but some recent attempts have taken place to identify categories of impacts and to understand advantages and limitations to their implementation in decision making (Hutchins, 2010). Parris and Kates (2003) reviewed 12 efforts to define indicators of sustainability, ranging in scale from global to local, and identified from 6 to 255 indicators. With such a large number of social sustainability indicators to work with, some attempt to organize them into themes and subthemes has proven beneficial. Table 5-6 lists several social indicator themes and subthemes developed by the United Nations for the purpose of tracking progress toward meeting the Millennium Development Goals (UNDSD, 2001).

Table 5-6   Selected UNDSD Indicators of Social Sustainable Development

Theme	Subtheme	Core Indicator
Poverty	Income poverty	Proportion of population living below national poverty line
	Income inequality	Ratio of share in national income of highest to lowest quintile
	Sanitation	Proportion of population using an improved sanitation facility
	Drinking water	Proportion of population using an improved water source
	Access to energy	Share of households without electricity or other modern energy services
	Living conditions	Proportion of urban population living in slums
Governance	Corruption	Percentage of population having paid bribes
	Crime	Number of intentional homicides per 100,000 population
Health	Mortality	Under-five mortality rate
	Health care delivery	Percent of population with access to primary health care facilities
	Nutritional status	Nutritional status of children
	Health status and risk	Morbidity of major diseases such as HIV/AIDS, malaria, tuberculosis
Education	Education level	Gross intake ratio to last grade primary education
		Net enrollment rate in primary education
		Adult secondary (tertiary) schooling attainment level
	Literacy	Adult literacy rate
Demographics	Population	Population, growth rate, dependency rate
Natural hazards	Vulnerability to natural hazards	Percentage of population living in hazard-prone areas

Source: UNDSD, 2001

Social Sustainability Categories							
Employee Working Conditions	Working Accidents	Fatal Working Accidents	Work - Related Diseases	Toxicity Potential / Transport	Wages and Salaries	Professional Training	Strikes and Lockouts
International Community	Child Labor	Foreign Direct Investment	Imports from Developing Countries				
Future Generations	Number of Trainees	R & D Spending	Capital Investment	Social Security			
Consumer	Toxicity Potential	Product Risks to Consumer					
Local & National Community	Employees	Qualified Employees	Gender Equity	Integration of Disabled People	Part-Time Workers	Family Support	

Figure 5-4 Social sustainability stakeholders (categories) and indicators within each category (Adapted from Kölsch et al., 2008)

Clearly these indicators are intended to measure progress toward meeting global sustainability objectives in the area of basic human needs. However, this set of indicators may not be the most appropriate for judging the performance of engineering designs. For example, it contains very few indicators that measure higher-level social aspirations in areas of higher education, research and development, and gender and generational equity.

Another recent effort to incorporate social indicators into product and process design was implemented into a life-cycle assessment framework (Kölsch et al., 2008). The goal of this methodology effort was to provide decision makers with a tool to judge the relative social costs and benefits alongside environmental life-cycle impacts and economic costs, achieving a comprehensive sustainability analysis. Figure 5-4 shows five categories of social impacts, the categories corresponding to key stakeholders for a global manufacturing corporation, and several indicators within these categories. There are similarities with the themes and subthemes from Table 5-6, but with additional indicators dealing with employment conditions, professional development, research, future generations, and gender equity. A case study comparing rapeseed biodiesel derived from plant oil to petroleum diesel based on social, environmental, and cost indicators was presented. Inventory data for the social sustainability analysis were assigned to key material and energy inputs into the fuel life cycles, similar to environmental life-cycle assessment, using national-level statistical data from major industry sectors. Thus, each key input to the product life cycle is accompanied by a detailed social sustainability profile. This method also discussed normalization and weighting approaches required to aggregate the numerous indicators in each stakeholder category into a single score, similar to environmental life-cycle assessment. The study concluded that petroleum diesel is preferable from an economic cost perspective and that biodiesel produced from rapeseed was superior based on both environmental and social life-cycle assessment.

As demonstrated in this section, it is possible to incorporate social sustainability into engineering design in a way that integrates with environmental life-cycle assessment and costs. The methods are rudimentary and there is little evidence to date that results are robust and accurate, but a framework for improvement has been achieved.

5.6 SUMMARY

Engineers play an important role in global sustainable development by developing and commercializing technologies that allow for a high standard of living, with favorable economic returns, minimum environmental impact, and with regard to the social conditions of stakeholders. This chapter provided engineering design principles to address environmental issues, introduced an environmental perspective on engineering economics, presented an overview of environmental indicators, and outlined a framework for social sustainability assessment. Taken together, these approaches, methods, and frameworks are a valuable addition to an engineer's design "toolbox."

PROBLEMS

1. **Design for a purpose** Provide several examples of engineered products that are flawed and wasteful when conforming to a "one size fits all" design. Do the same for products that are efficient because they are defined for a specific purpose.

2. **Improving business performance through improving environmental performance** Browse the Web site of the World Business Council for Sustainable Development (www.wbcsd.ch) and identify a case study of a company improving business performance through improving environmental performance. Write a one-page summary of the case study.

3. **Sustainability reports** Conduct an Internet search to obtain a recent sustainability report for a major corporation. Compare reported environmental costs with total revenues. Determine the percent of total revenues devoted to environmental expenditures and compare it to the percentages listed in Table 5-1. Discuss any itemization of environmental expenditures in the report (environmental operations, capital equipment, remediation, reporting, etc.).

4. **Risk shifting** Provide examples of "risk shifting," a concept introduced in Sandestin Green Engineering Principle 1.

5. **Unintended consequences** Provide examples of engineering designs or products from these designs that caused unintended damage to ecosystems (Sandestin Green Engineering Principle 2).

6. **Life cycles** Sketch the life cycle of a product of engineering design. Include inputs and outputs of key materials and energy at each stage of the life cycle, including recycle/disposal at end of life as well as transportation between stages (Sandestin Green Engineering Principle 3).

7. **Safety** Discuss the inherent safety issues of petroleum versus forest biomass as feed-stock for production of liquid transportation fuels. Both of these feedstocks can be converted to liquid hydrocarbon fuels. Consider the safety to both human health and ecosystems in your answer (Sandestin Green Engineering Principle 4).

8. **Natural resource use** What natural resources are under- or overutilized in your area of the United States? How could engineering designs address resource scarcity in such cases (Sandestin Green Engineering Principle 5)?

9. **Wastes as raw materials** Bagasse, a waste product of sugarcane processing, is the woody stalk of the cane plant remaining after sugar is extracted. In Brazil, for example, bagasse is often used as a fuel for power generation at the locations of cane processing. Provide one or more examples of wastes from key product systems that have found beneficial uses through engineering designs (Sandestin Green Engineering Principle 6).

10. **Personal transportation** Investigate differences in personal transportation from country to country and elaborate on cultural contexts of these different transportation options (Sandestin Green Engineering Principle 7).

11. **Engineering for sustainability** How have innovations in engineering design helped achieve sustainability goals in developing countries, such as providing basic services like drinking water, power, food preservation, and sanitation? Provide some examples and discuss the engineering design aspects (Sandestin Green Engineering Principle 8).

12. **Community engagement** Investigate and report on how engineers became involved in community engagement in your chosen field of study. What kinds of community engagement activities were the engineers involved in (Sandestin Green Engineering Principle 9)?

13. **Global warming emissions from electricity generation** Each state in the United States has a unique profile of electricity generation types, and this characteristic is also true for cities within these states. Using the table of electricity generation sources below:
    a. Calculate in a table the global warming index for each city's electricity based on 1 kWh generated.
    b. Compare and discuss the global warming index for each city. Which city has the lowest global warming index? Which city has the highest index? By how much do the high and low cities differ (percent of the highest)?
    Data for cities are from the U.S. EPA at www.epa.gov/cleanenergy/energy-and-you/index.html. This problem was originally analyzed by Mark Cleghorn, an online student at Michigan Technological University, in the course ENG5510 Sustainable Futures 1.

| | Electricity (Total of 1 kWh) | | | | | | | |
City	Unit	Hard Coal	Oil	Natural Gas	Nuclear	Hydro	Wind	PV[a]
Dallas	kWh	0.371	0.005	0.475	0.119	0.003	0.014	0.013
Detroit	kWh	0.669	0.008	0.137	0.156	0	0.011	0.019
Los Angeles	kWh	0.119	0.012	0.483	0.165	0.117	0.01	0.094
Phoenix	kWh	0.457	0.001	0.316	0.164	0.035	0	0.026
Pittsburgh	kWh	0.728	0.004	0.027	0.223	0.007	0.007	0.004
Seattle	kWh	0.344	0.003	0.108	0.033	0.486	0.003	0.023
U.S. average[b]	kWh	0.445	0.01	0.233	0.202	0.069	0.019	0.018

[a]PV = photovoltaic but is meant to represent all other renewable electricity besides wind and hydro.
[b]Data from the U.S. Department of Energy, www.eia.doe.gov/cneaf/electricity/epa/epa_sum.html.

APPENDIX

Table A5-1 Global Warming Potentials for Greenhouse Gases (CO_2 Is the Benchmark)

Industrial Designation or Common Name (yrs)	Chemical Formula	Lifetime (yrs)	Radiative Efficiency (W m^{-2} ppb^{-1})	Global Warming Potential for Given Time Horizon			
				SARa (100-yr)	20-yr	100-yr	500-yr
Carbon dioxide	CO_2	See below[b]	$1.4 \times 10^{-5\,c}$	1	1	1	1
Methane[d]	CH_4	12[d]	3.7×10^{-4}	21	72	25	7.6
Nitrous oxide	N_2O	114	3.03×10^{-3}	310	289	298	153
Substances Controlled by the Montreal Protocol							
CFC-11	CCl_3F	45	0.25	3800	6730	4750	1620
CFC-12	CCl_2F_2	100	0.32	8100	11,000	10,900	5200
CFC-13	$CClF_3$	640	0.25		10,800	14,400	16,400
CFC-113	CCl_2FCClF_2	85	0.3	4800	6540	6130	2700
CFC-114	$CClF_2CClF_2$	300	0.31		8040	10,000	8730
CFC-115	$CClF_2CF_3$	1700	0.18		5310	7370	9990
Halon-1301	$CBrF_3$	65	0.32	5400	8480	7140	2760
Halon-1211	$CBrClF_2$	16	0.3		4750	1890	575
Halon-2402	$CBrF_2CBrF_2$	20	0.33		3680	1640	503
Carbon tetrachloride	CCl_4	26	0.13	1400	2700	1400	435
Methyl bromide	CH_3Br	0.7	0.01		17	5	1
Methyl chloroform	CH_3CCl_3	5	0.06		506	146	45
HCFC-22	$CHClF_2$	12	0.2	1500	5160	1810	549
HCFC-123	$CHCl_2CF_3$	1.3	0.14	90	273	77	24

(*Continued*)

Table A5-1 (*Continued*)

Industrial Designation or Common Name (yrs)	Chemical Formula	Lifetime (yrs)	Radiative Efficiency (W m^{-2} ppb^{-1})	Global Warming Potential for Given Time Horizon			
				SAR[a] (100-yr)	20-yr	100-yr	500-yr
HCFC-124	CHClFCF$_3$	5.8	0.22	470	2070	609	185
HCFC-141b	CH$_3$CCl$_2$F	9.3	0.14		2250	725	220
HCFC-142b	CH$_3$CClF$_2$	17.9	0.2	1800	5490	2310	705
HCFC-225ca	CHCl$_2$CF$_2$CF$_3$	1.9	0.2		429	122	37
HCFC-225cb	CHClFCF$_2$CClF$_2$	5.8	0.32		2030	595	181
Hydrofluorocarbons							
HFC-23	CHF$_3$	270	0.19	11,700	12,000	14,800	12,200
HFC-32	CH$_2$F$_2$	4.9	0.11	650	2330	675	205
HFC-125	CHF$_2$CF$_3$	29	0.23	2800	6350	3500	1100
HFC-134a	CH$_2$FCF$_3$	14	0.16	1300	3830	1430	435
HFC-143a	CH$_3$CF$_3$	52	0.13	3800	5890	4470	1590
HFC-152a	CH$_3$CHF$_2$	1.4	0.09	140	437	124	38
HFC-227ea	CF$_3$CHFCF$_3$	34.2	0.26	2900	5310	3220	1040
HFC-236fa	CF$_3$CH$_2$CF$_3$	240	0.28	6300	8100	9810	7660
HFC-245fa	CHF$_2$CH$_2$CF$_3$	7.6	0.28		3380	1030	314
HFC-365mfc	CH$_3$CF$_2$CH$_2$CF$_3$	8.6	0.21		2520	794	241
HFC-43-10mee	CF$_3$CHFCHFCF$_2$CF$_3$	15.9	0.4	1300	4140	1640	500
Perfluorinated Compounds							
Sulfur hexafluoride	SF$_6$	3200	0.52	23,900	16,300	22,800	32,600
Nitrogen trifluoride	NF$_3$	740	0.21		12,300	17,200	20,700
PFC-14	CF$_4$	50,000	0.10	6500	5210	7390	11,200
PFC-116	C$_2$F$_6$	10,000	0.26	9200	8630	12,200	18,200
PFC-218	C$_3$F$_8$	2600	0.26	7000	6310	8830	12,500
PFC-318	c-C$_4$F$_8$	3200	0.32	8700	7310	10,300	14,700
PFC-3-1-10	C$_4$F$_{10}$	2600	0.33	7000	6330	8860	12,500
PFC-4-1-12	C$_5$F$_{12}$	4100	0.41		6510	9160	13,300
PFC-5-1-14	C$_6$F$_{14}$	3200	0.49	7400	6600	9300	13,300
PFC-9-1-18	C$_{10}$F$_{18}$	>1000[e]	0.56		>5500	>7500	>9500
Trifluoromethyl sulfur pentafluoride	SF$_5$CF$_3$	800	0.57		13,200	17,700	21,200
Fluorinated Ethers							
HFE-125	CHF$_2$OCF$_3$	136	0.44		13,800	14,900	8490
HFE-134	CHF$_2$OCHF$_2$	26	0.45		12,200	6320	1960

Table A5-1   (*Continued*)

Industrial Designation or Common Name (yrs)	Chemical Formula	Lifetime (yrs)	Radiative Efficiency (W m^{-2} ppb^{-1})	Global Warming Potential for Given Time Horizon			
				SAR^a (100-yr)	20-yr	100-yr	500-yr
HFE-143a	CH_3OCF_3	4.3	0.27		2630	756	230
HCFE-235da2	$CHF_2OCHClCF_3$	2.6	0.38		1230	350	106
HFE-245cb2	$CH_3OCF_2CHF_2$	5.1	0.32		2440	708	215
HFE-245fa2	$CHF_2OCH_2CF_3$	4.9	0.31		2280	659	200
HFE-254cb2	$CH_3OCF_2CHF_2$	2.6	0.28		1260	359	109
HFE-347mcc3	$CH_3OCF_2CF_2CF_3$	5.2	0.34		1980	575	175
HFE-347pcf2	$CHF_2CF_2OCH_2CF_3$	7.1	0.25		1900	580	175
HFE-356pcc3	$CH_3OCF_2CF_2CHF_2$	0.33	0.93		386	110	33
HFE-449sl (HFE-7100)	$C4F_9OCH_3$	3.8	0.31		1040	297	90
HFE-569sf2 (HFE-7200)	$C_4F_9OC_2H_5$	0.77	0.3		207	59	18
HFE-43-10pccc124 (H-Galden 1040x)	$CHF_2OCF_2OC_2$- F_4OCHF_2	6.3	1.37		6320	1870	569
HFE-236ca12 (HG-10)	$CHF_2OCF_2OCHF_2$	12.1	0.66		8000	2800	860
HFE-338pcc13 (HG-01)	$CHF_2OCF_2CF_2O$- CHF_2	6.2	0.87		5100	1500	460
Perfluoropolyethers							
PFPMIE	$CF_3OCF(CF_3)$ $CF_2OCF_2OCF_2$	800	0.65		7620	10,300	12,400
Hydrocarbons and Other Compounds — Direct Effects							
Dimethylether	CH_3OCH_3	0.015	0.02		1	1	<<1
Methylene chloride	CH_2Cl_2	0.38	0.03		31	8.7	2.7
Methyl chloride	CH_3Cl	1.0	0.01		45	13	4

^{a} SAR refers to the IPCC *Second Assessment Report* (1995) used for reporting under the UNFCCC.

^{b} The CO_2 response function is based on the revised version of the Bern Carbon cycle model using a background CO_2 concentration value of 378 ppm. The decay of a pulse of CO_2 with time t is given by

$$a_0 + \sum_{j=1}^{3} a_j e \frac{-t}{\tau_j} \text{ where } a_0 = 0.217, a_1 = 0.259, a_2 = 0.338, a_3 = 0.186, \tau_1 = 172.9 \text{ yrs}, \tau_2 = 18.51 \text{ yrs},$$

and $\tau_3 = 1.186$ yrs, for t < 1000 yrs.

^{c} The radiative efficiency of CO_2 is calculated using the simplified expression as revised in the TAR, with an updated background concentration value of 378 ppm and a perturbation of +1 ppm.

^{d} The perturbation lifetime for CH_4 is 12 years as in the TAR. The GWP for CH_4 includes indirect effects from enhancements of ozone and stratospheric water vapor.

^{e} The assumed lifetime of 1000 years is a lower limit.

Source: IPCC, 2007

Table A5-2   Ozone Depletion Potentials for Class I Substances to Be Phased Out

Chemical Name or Formula	Life-time (yrs)	ODP3 (WMO 2006)	ODP2 (40 CFR 82)	ODP1 (Montreal Protocol)	CAS Number
Group I (from Section 602 of the Clean Air Act)					
CFC-11 (CCl_3F) trichlorofluoromethane	45	1	1	1	75-69-4
CFC-12 (CCl_2F_2) dichlorodifluoromethane	100	1	1	1	75-71-8
CFC-113 ($C_2F_3Cl_3$) 1,1,2-trichlorotrifluoro-ethane	85	1	0.8	0.8	76-13-1
CFC-114 ($C_2F_4Cl_2$) dichlorotetrafluoroethane	300	1	1	1	76-14-2
CFC-115 (C_2F_5Cl) monochloropentafluoro-ethane	1700	0.44	0.6	0.6	76-15-3
Group II (from Section 602 of the Clean Air Act)					
Halon 1211 (CF_2ClBr) bromochlorodifluo-romethane	16	7.1	3	3	353-59-3
Halon 1301 (CF_3Br) bromotrifluoromethane	65	16	10	10	75-63-8
Halon 2402 ($C_2F_4Br_2$) dibromotetrafluoro-ethane	20	11.5	6	6	124-73-2
Group III (from Section 602 of the Clean Air Act)					
CFC-13 (CF_3Cl) chlorotrifluoromethane	640		1	1	75-72-9
CFC-111 (C_2FCl_5) pentachlorofluoroethane			1	1	354-56-3
CFC-112 ($C_2F_2Cl_4$) tetrachlorodifluoroethane			1	1	76-12-0
CFC-211 (C_3FCl_7) heptachlorofluoropropane			1	1	422-78-6
CFC-212 ($C_3F_2Cl_6$) hexachlorodifluoropropane			1	1	3182-26-1
CFC-213 ($C_3F_3Cl_5$) pentachlorotrifluoropropane			1	1	6/5/2354
CFC-214 ($C_3F_4Cl_4$) tetrachlorotetrafluoro-propane			1	1	29255-31-0
CFC-215 ($C_3F_5Cl_3$) trichloropentafluoropropane			1	1	4259-43-2
CFC-216 ($C_3F_6Cl_2$) dichlorohexafluoropropane			1	1	661-97-2
CFC-217 (C_3F_7Cl) chloroheptafluoropropane			1	1	422-86-6
Group IV (from Section 602 of the Clean Air Act)					
CCl_4 carbon tetrachloride	26	0.73	1.1	1.1	56-23-5
Group V (from Section 602 of the Clean Air Act)					
Methyl chloroform ($C_2H_3Cl_3$) 1,1,1-trichloro-ethane	5	0.12	0.1	0.1	71-55-6
Group VI (Listed in the Accelerated Phaseout Final Rule)					
Methyl bromide (CH_3Br)	0.7	0.51	0.7	0.6	74-83-9
Group VII (Listed in the Accelerated Phaseout Final Rule)					
$CHFBr_2$			1	1	
HBFC-12B1(CHF_2Br)				0.74	
CH_2FBr			0.73	0.73	

Table A5-2 *(Continued)*

Chemical Name or Formula	Life-time (yrs)	ODP3 (WMO 2006)	ODP2 (40 CFR 82)	ODP1 (Montreal Protocol)	CAS Number
C_2HFBr_4			0.3–0.8	0.3–0.8	
$C_2HF_2Br_3$			0.5–1.8	0.5–1.8	
$C_2HF_3Br_2$			0.4–1.6	0.4–1.6	
C_2HF_4Br			0.7–1.2	0.7–1.2	
$C_2H_2FBr_3$			0.1–1.1	0.1–1.1	
$C_2H_2F_2Br_2$			0.2–1.5	0.2–1.5	
$C_2H_2F_3Br$			0.7–1.6	0.7–1.6	
$C_2H_3FBr_2$			0.1–1.7	0.1–1.7	
$C_2H_3F_2Br$			0.2–1.1	0.2–1.1	
C_2H_4FBr			0.07–0.1	0.07–0.1	
C_3HFBr_6			0.3–1.5	0.3–1.5	
$C_3HF_2Br_5$			0.2–1.9	0.2–1.9	
$C_3HF_3Br_4$			0.3–1.8	0.3–1.8	
$C_3HF_4Br_3$			0.5–2.2	0.5–2.2	
$C_3HF_5Br_2$			0.9–2.0	0.9–2.0	
C_3HF_6Br			0.7–3.3	0.7–3.3	
$C_3H_2F_2Br_4$			0.2–2.1	0.2–2.1	
$C_3H_2F_3Br_3$			0.2–5.6	0.2–5.6	
$C_3H_2F_4Br_2$			0.3–7.5	0.3–7.5	
$C_3H_2F_5Br$			0.9–14	0.9–1.4	
$C_3H_3FBr_4$			0.08–1.9	0.08–1.9	
$C_3H_3F_2Br_3$			0.1–3.1	0.1–3.1	
$C_3H_3F_3Br_2$			0.1–2.5	0.1–2.5	
$C_3H_3F_4Br$			0.3–4.4	0.3–4.4	
$C_3H_4FBr_3$			0.03–0.3	0.03–0.3	
$C_3H_4F_2Br_2$			0.1–1.0	0.1–1.0	
$C_3H_4F_3Br$			0.07–0.8	0.07–0.8	
$C_3H_5FBr_2$			0.04–0.4	0.04–0.4	
$C_3H_5F_2Br$			0.07–0.8	0.07–0.8	
C_3H_6FBr			0.02–0.7	0.02–0.7	
Group VIII (from the Chlorobromomethane Phaseout Final Rule)					
CH_2BrCl chlorobromomethane	0.37		0.12	0.12	

Source: EPA, 2011

Table A5-3 Ozone Depletion Potentials for Class II Substances—Substitutes

Chemical Name or Formula	Lifetime (yrs)	ODP3 (WMO 2006)	ODP2 (40 CFR 82)	ODP1 (Montreal Protocol)	CAS Number
HCFC-21 ($CHFCl_2$) dichlorofluoromethane	1.7		0.04	0.04	75-43-4
HCFC-22 (CHF_2Cl) monochlorodifluoromethane	12	0.05	0.055	0.055	75-45-6
HCFC-31 (CH_2FCl) monochlorofluoromethane			0.02	0.02	593-70-4
HCFC-121 (C_2HFCl_4) tetrachlorofluoroethane			0.01–0.04	0.01–0.04	354-14-3
HCFC-122 ($C_2HF_2Cl_3$) trichlorodifluoroethane			0.02–0.08	0.02–0.08	354-21-2
HCFC-123 ($C_2HF_3Cl_2$) dichlorotrifluoroethane	1.3	0.02	0.02	0.02	306-83-2
HCFC-124 (C_2HF_4Cl) monochlorotetrafluoroethane	5.8	0.022	0.022	0.022	2837-89-0
HCFC-131 ($C_2H_2FCl_3$) trichlorofluoroethane			0.007–0.05	0.007–0.05	359-28-4
HCFC-132b ($C_2H_2F_2Cl_2$) dichlorodifluoroethane			0.008–0.05	1649-08-7
HCFC-133a ($C_2H_2F_3Cl$) monochlorotrifluoroethane				0.02–0.06	75-88-7
HCFC-141b ($C_2H_3FCl_2$) dichlorofluoroethane	9.3	0.12	0.11	0.11	1717-00-6
HCFC-142b ($C_2H_3F_2Cl$) monochlorodifluoroethane	17.9	0.07	0.065	0.065	75-68-3
HCFC-221 (C_3HFCl_6) hexachlorofluoropropane			0.015–0.07	0.015–0.07	422-26-4
HCFC-222 ($C_3HF_2Cl_5$) pentachlorodifluoropropane			0.01–0.09	0.01–0.09	422-49-1
HCFC-223 ($C_3HF_3Cl_4$) tetrachlorotrifluoropropane			0.01–0.08	0.01–0.08	422-52-6
HCFC-224 ($C_3HF_4Cl_3$) trichlorotetrafluoropropane			0.01–0.09	0.01–0.09	422-54-8
HCFC-225ca ($C_3HF_5Cl_2$) dichloropentafluoropropane	1.9	0.02	0.025	0.025	422-56-0
HCFC-225cb ($C_3HF_5Cl_2$) dichloropentafluoropropane	5.8	0.03	0.033	0.033	507-55-1
HCFC-226 (C_3HF_6Cl) monochlorohexafluoropropane			0.02–0.1	0.02–0.1	431-87-8
HCFC-231 ($C_3H_2FCl_5$) pentachlorofluoropropane			0.05–0.09	0.05–0.09	421-94-3
HCFC-232 ($C_3H_2F_2Cl_4$) tetrachlorodifluoropropane			0.008–0.1	0.008–0.1	460-89-9

Table A5-3 (*Continued*)

Chemical Name or Formula	Lifetime (yrs)	ODP3 (WMO 2006)	ODP2 (40 CFR 82)	ODP1 (Montreal Protocol)	CAS Number
HCFC-233 ($C_3H_2F_3Cl_3$) trichlorotrifluoropropane			0.007–0.23	0.007–0.23	7125-84-0
HCFC-234 ($C_3H_2F_4Cl_2$) dichlorotetrafluoropropane			0.01–0.28	0.01– - 0.28	425-94-5
HCFC-235 ($C_3H_2F_5Cl$) monochloropentafluoropropane			0.03–0.52	0.03–0.52	460-92-4
HCFC-241 ($C_3H_3FCl_4$) tetrachlorofluoropropane			0.004–0.09	0.004–0.09	666-27-3
HCFC-242 ($C_3H_3F_2Cl_3$) trichlorodifluoropropane			0.005–0.13	0.005–0.13	460-63-9
HCFC-243 ($C_3H_3F_3Cl_2$) dichlorotrifluoropropane			0.007–0.12	0.007–0.12	460-69-5
HCFC-244 ($C_3H_3F_4Cl$) monochlorotetrafluoropropane			0.009–0.14	0.009–0.14	
HCFC-251 ($C_3H_4FCl_3$) monochlorotetrafluoropropane			0.001–0.01	0.001–0.01	421-41-0
HCFC-252 ($C_3H_4F_2Cl_2$) dichlorodifluoropropane			0.005–0.04	0.005–0.04	819-00-1
HCFC-253 ($C_3H_4F_3Cl$) monochlorotrifluoropropane			0.003–0.03	0.003–0.03	460-35-5
HCFC-261 ($C_3H_5FCl_2$) dichlorofluoropropane			0.002–0.02	0.002–0.02	420-97-3
HCFC-262 ($C_3H_5F_2Cl$) monochlorodifluoropropane			0.002–0.02	0.002–0.02	421-02-03
HCFC-271 (C_3H_6FCl) monochlorofluoropropane			0.001–0.03	0.001–0.03	430-55-7

Source: EPA, 2011

Table A5-4 Acid Rain Potential for a Number of Acidifying Chemicals

Compound		α	MW_i (mol/kg)	η_{i}, (mol H^+/ kg i)	ARP_i
SO_2	$SO_2 + H_2O + O_3 \rightarrow 2H^+ + SO_4^{2-} + O_2$	2	0.064	31.25	1.00
NO	$SO + O_3 + 1/2\ H_2O \rightarrow H^+ + NO_3^- + 3/4\ O_2$	1	0.030	33.33	1.07
NO_2	$NO_2 + 1/2\ H_2O + 1/4\ O_2 \rightarrow H^+ + NO_3^-$	1	0.046	21.74	0.70
NH_3	$NH_3 + 2\ O_2 \rightarrow H^+ + NO_3^- + H_2O$	1	0.017	58.82	1.88
HCl	$HCl \rightarrow H^+ + Cl^-$	1	0.0365	27.40	0.88
HF	$HF \rightarrow H^+ + F^-$	1	0.020	50.00	1.60

Source: Adapted from Heijungs et al., 1992

Table A5-5 Maximum Incremental Reactivities (MIR; g O_3/g VOC) for Smog Formation (Tropospheric O_3)

Alkanes	normal	MIR	branched	MIR
	methane	0.0144	isobutane	0.98
	ethane	0.28	neopentane	1.72
	propane	0.49	iso-pentane	2.09
	n-butane	1.15	2,2-dimethylbutane	0.77
	n-pentane	1.31	2,3-dimethylpentane	0.66
	n-hexane	1.24	2-methylpentane	10.56
	n-heptane	1.07	3-methylpentane	4.62
	n-octane	0.90	2,2,3-trimethylbutane	0.55
	n-nonane	0.78	2,4-dimethylpentane	0.66
	n-decane	0.68	3,3-dimethylpentane	6.61
	n-undecane	1.54	2-methylhexane	7.21
	n-dodecane	1.28	3-methylhexane	4.54
	n-tridecane	1.84	2,2,4-trimethylpentane	0.70
	n-tetradecane	2.32	2,3,4-trimethylpentane	0.47
	Average	**1.06**	2,3-dimethylhexane	0.60
			2,4-dimethylhexane	0.50
	cyclic		2,5-dimethylhexane	0.43
	cyclopentane	7.09	2-methylheptane	6.99
	methylcyclopentane	0.80	3-methylheptane	3.13
	cyclohexane	1.15	4-methylheptane	2.59
	1,3-dimethylcyclohexane	0.78	2,4-dimethylheptane	0.87
	methylcyclohexane	0.55	2,2,5-trimethylhexane	0.80
	ethylcyclopentane	1.84	4-ethylheptane	2.39
	ethylcyclohexane	2.55	3,4-propylheptane	1.01
	1-ethyl-4-methylcyclohexane	1.36		
	1,3-diethylcyclohexane	1.46	3,5-diethylheptane	9.75
	1,3-diethyl-5-methylcyclohexane	1.47	2,6-diethyloctane	0.52
	1,3,5-triethylcyclohexane	1.01	**Average**	**2.68**
	Average	**1.82**		
Alkenes	primary		secondary	
	ethene	2.40	isobutene	1.07
	propene	0.28	2-methyl-1-butene	6.77
	1-butene	0.99	trans-2-butene	1.17
	1-pentene	1.23	cis-2-butene	15.61
	3-methyl-1-butene	3.66	2-pentenes	8.62
	1-hexene	1.28	2-methyl-2-butene	10.29
	1-heptene	0.68	2-hexenes	14.24
	1-octene	1.07	2-heptenes	9.73
	1-nonene	1.09	3-octenes	3.31
	Average	**1.40**	3-nonenes	4.51
			Average	**7.53**

Table A5-5   (*Continued*)

Alkanes	*others*	MIR		MIR
	1,3-butadiene	1.64	**Alcohols and Ethers**	
	isoprene	2.58	methanol	0.92
	cyclopentene	1.35	ethanol	2.40
	cyclohexene	2.66	n-propyl alcohol	1.25
	α-pin*ene*	6.23	isopropyl alcohol	1.62
	β-pinene	5.49	n-butyl alcohol	1.33
	Average	**3.33**	isobutyl alcohol	1.08
			t-butyl alcohol	−0.13
Acetylenes			dimethyl ether	0.60
	acetylene	2.52	methyl t-butyl ether	1.38
	methylacetylene	0.65	ethyl t-butyl ether	1.09
	Average	**1.59**	**Average**	**1.15**
Aromatics				
	benzene	2.36	**Aromatic Oxygenates**	
	toluene	2.70	benzaldehyde	2.36
	ethylbenzene	1.53	phenol	3.70
	n-propylbenzene	0.38	alkyl phenols	2.30
	isopropylbenzene	1.15	**Average**	**2.79**
	s-butylbenzene	0.00		
	o-xylene	1.79		
	p-xylene	5.62	**Aldehydes**	
	m-xylene	0.61	formaldehyde	1.98
	1,3,5-trimethylbenzene	1.22	acetaldehyde	7.64
	1,2,3-trimethylbenzene	1.11	C3 aldehydes	6.50
	1,2,4-trimethylbenzene	1.70	glyoxal	2.12
	tetralin	0.29	methyl glyoxal	2.61
	naphthalene	1.17	**Average**	**4.17**
	methylnaphthalenes	0.86		
	2,3-dimethylnaphthalene	0.51		
	styrene	3.46		
	Average	**1.56**	**Ketones**	
			acetone	2.03
			C4 ketones	1.18
Others			**Average**	**1.61**
	Methyl nitrite	0.61		
	Base Reactive Organic Gas Mixture			3.10

Source: From R. P. L. Carter, University of California, Riverside, www.engr.ucr.edu/~carter/SAPRC/

REFERENCES

Abraham, M. A., ed. 2006. *Sustainability Science and Engineering: Defining Principles.* Amsterdam: Elsevier Science.

Abraham, M. A., and N. Nguyen. 2003. "Green Engineering: Defining the Principles." Results from the Sandestin Conference. *Environmental Progress* 22(4):233–36.

Allen, D. T., N. Bakshani, and K. S. Rosselot. 1992. *Pollution Prevention: Homework & Design Problems for Engineering Curricula.* New York: American Institute of Chemical Engineers.

Allen, D. T., and K. S. Rosselot. 1997. *Pollution Prevention for Chemical Processes,* 1st ed. New York: John Wiley & Sons.

Allen, D. T., and D. R. Shonnard (and other contributors). *Green Engineering: Environmentally Conscious Design of Chemical Processes,* 2nd ed. Upper Saddle River, NJ: Prentice Hall (In press).

AIChE CWRT (American Institute of Chemical Engineers' Center for Waste Reduction Technologies). 2000. *Total Cost Assessment Methodology.* New York: AIChE.

Anastas, Paul T., and John C. Warner. "Tweve Principles of Green Chemistry." Available at www.chm.bris.ac.uk/webprojects2002/howells/twelve_green_chemistry_principle.htm.

Anastas, Paul T., and J. B. Zimmerman. 2003. "Design Through the 12 Principles of Green Engineering." *Environmental Science & Technology* 37(5):94A–l01A.

Bare, J. C., P. Hofstetter, D. W. Pennington, and H. A. Udo de Haes. 2000. "Midpoints vs Endpoints: The Sacrifices and Benefits." *International Journal of Life Cycle Assessment* 5(6):319–26.

Bare, J. C., G. A. Norris, D. W. Pennington, and T. McKone. 2003. "TRACI: The Tool for the Reduction and Assessment of Chemical and Other Environmental Impacts." *Journal of Industrial Ecology* 6(3–4):49–78.

Boustead, I. 1993. "Ecoprofiles of the European Plastics Industry, Report 1–4." Reports for the European Centre for Plastics in the Environment (PWMI), Brussels, May.

Carter, W. P., and R. Atkinson. 1989. "A Computer Modeling Study of Incremental Hydrocarbon Reactivity." *Environmental Science & Technology* 23:864–80.

Carter, W. P. L. 1994. "Development of Ozone Reactivity Scales for Volatile Organic Compounds. *Air & Waste* 44:881–99.

Chang, T. Y., and S. J. Rudy. 1990. "Ozone-Forming Potential of Organic Emissions from Alternative-Fueled Vehicles." *Atmospheric Environment* 24A:2421.

Derwent, R. G. 1990. *Trace Gases and Their Relative Contribution to the Greenhouse Effect.* Report AERE-R13716. Harwell, UK: Atomic Energy Research Establishment.

Dodge, M. C. 1984. "Combined Effects of Organic Reactivity and NMHC/NOx Ratio on Photochemical Oxidant Formation: A Modeling Study." *Atmospheric Environment* 18:1857.

Fava, J., et al. 1992. *Life Cycle Assessment Data Quality: A Conceptual Framework.* SETAC workshop, Wintergreen, VA.

Fava, J., and F. Consoli. 1996. "Application of Life-Cycle Assessment to Business Performance." In *Environmental Life Cycle Assessment,* edited by M. A. Curran. New York: McGraw-Hill.

Financial Accounting Standards Board. 1985. Concept Statement No. 6, Paragraph 35.

Fisher, D. A., C. H. Hales, W. Wang, M. K. W. Ko, and N. D. Sze. 1990a. "Model Calculations of the Relative Effects of CFCs and Their Replacements on Global Warming." *Nature* 344:513–16.

Fisher, D. A., C. H. Hales, D. L. Filkin, M. K. W. Ko, N. D. Sze, P. S. Connell, D. J. Wuebbles, I. S. A. Isaksen, and F. Stordal. 1990b. "Model Calculations of the Relative Effects of CFCs and Their Replacements on Stratospheric Ozone." *Nature* 344:508–12.

Goedkoop, M., and R. Spriensma. 1999. *The Eco-Indicator'99: A Damage-Oriented Method for Life Cycle Impact Assessment.* Zoetermeer, The Netherlands: VROM, Ministry of Housing, Spatial Planning, and the Environment. www.pre.nl.

Graedel, T. E., and B. R. Allenby. 1998. *Industrial Ecology and the Automobile.* Upper Saddle River, NJ: Prentice Hall.

Guinée, J., R. Heijungs, L. van Oers, D. van de Meent, T. Vermeire, and M. Rikken. 1996. *LCA Impact Assessment of Toxic Releases: Generic Modelling of Fate, Exposure, and Effect for Ecosystems and Human Beings with Data for About 100 Chemicals.* No. 1996/21. The Hague, The Netherlands: VROM, Ministry of Housing, Spatial Planning, and the Environment.

Heijungs, R., J. B. Guinée, G. Huppes, R. M. Lankreijer, H. A. Udo de Haes, and A. Wegener Sleeswijk. 1992. "Environmental Life Cycle Assessment of Products. Guide and Backgrounds." NOH Report Numbers 9266 and 9267. Netherlands Agency for Energy and the Environment. November.

Heller, M., P. D. Shields, and B. Beloff. 1995. "Environmental Accounting Case Study: Amoco Yorktown Refinery." In *Green Ledgers: Case Studies in Corporate Environmental Accounting*, edited by D. Ditz, J. Ranganathan, and D. Banks. Washington, DC: World Resources.

Hertwich, E. G., S. F. Mateles, W. S. Pease, and T. E. McKone. 2001. "Human Toxicity Potentials for Life Cycle Analysis and Toxics Release Inventory Risk Screening." *Environmental Toxicology and Chemistry* 20:928–39.

Hutchins, M. 2010. *Framework, Indicators, and Techniques to Support Decision Making Related to Societal Sustainability.* Ph.D. Dissertation, Department of Mechanical Engineering, Michigan Technological University.

IISD (International Institute for Sustainable Development). 2002. *Seven Questions to Sustainability.* Task 2 Work Group, Mining, Minerals and Sustainable Development North America Project. www.iisd.org/publications/pub.aspx?id=456.

Institute of Management Accountants. 1990. *Management Accounting Glossary.* Statement No. 2A.

IPCC (Intergovernmental Panel on Climate Change). 2007. *Climate Change 2007: The Physical Science Basis*, edited by S. Solomon, D. Qin, and M. Manning. Cambridge: Cambridge University Press.

Japar, S. M., T. J. Wallington, S. J. Rudy, and T. Y. Chang. 1991. "Ozone-Forming Potential of a Series of Oxygenated Organic Compounds." *Environmental Science & Technology* 25:415–20.

Kölsch, D., P. Saling, A. Kicherer, A. Grosse-Sommer, and I. Schmidt. 2008. "How to Measure Social Impacts? A Socio-eco-efficiency Analysis by the SEEBALANCE® Method." *International Journal of Sustainable Development* 11(1):1–23.

Landsiedel, R., and P. Saling. 2002. "Assessment of Toxicological Risks for Lifecycle Assessment and Eco-Efficiency Analysis." *International Journal of Lifecycle Assessment* 7(5):261–68.

Lashof, D. A., and D. R. Ahuja. 1990. "Relative Contributions of Greenhouse Gas Emissions to Global Warming." *Nature* 344:529–31.

McDonough, W., and M. Braungart. 1992. "Hannover Principles." www.mindfully.org/Sustainability/Hannover-Principles.htm.

NRC (National Research Council), Committee on Institutional Means for Assessment of Risks to Public Health. 1983. *Risk Assessment in the Federal Government: Managing the Process.* Washington, DC: National Academy Press.

Parris, T. M., and R. W. Kates. 2003. "Characterizing and Measuring Sustainable Development." *Annual Review of Environment and Resources* 28: 559–86.

Postlethwaite, D., and N. de Oude. 1996. "European Perspective." Chapter 9 in *Environmental Life-Cycle Assessment*, edited by M. A. Curran. New York: McGraw-Hill.

Postlethwaite, D., B. Quay, J. Seguin, and B. Vigon, eds. 1993. *Guidelines for Life Cycle Assessment: A Code of Practice*. Pensacola, FL: SETAC Press.

Pratt, G. C., P. E. Gerbec, S. K. Livingston, F. Oliaei, G. L. Gollweg, S. Paterson, and D. Mackay. 1993. "An Indexing System for Comparing Toxic Air Pollutants Based upon Their Potential Environmental Impacts." *Chemosphere* 27(8):1350–79.

Ravishankara, A. R., A. A. Turnipseed, N. R. Jensen, S. Barone, M. Mills, C. J. Howard, and S. Solomon. 1994. "Do Hydrofluorocarbons Destroy Stratospheric Ozone?" *Science* 263:71–75.

Rosselot, K. S., and D. T. Allen. 1999. "Chlorinated Organic Compounds in the Great Lakes Basin: Impact Assessment." Submitted to *Journal of Industrial Ecology*.

Rotmans, J. 1990. *IMAGE: An Integrated Model to Assess the Greenhouse Effect*. Maastrich, Dordrecht, The Netherlands: Kluwer Academic Publishers, pp. 205–24.

Saling, P., A. Kicherer, B. Dittrich-Krämer, R. Wittlinger, W. Zombik, I. Schmidt, W. Schrott, and S. Schmidt. 2002. "Eco-efficiency Analysis by BASF: The Method." *International Journal of LCA* 7(4):203.

SETAC (Society for Environmental Toxicology and Chemistry). 1993. *Guidelines for Life-Cycle Assessment: Code of Practice*. Brussels, Belgium: SETAC Press.

Shields, P., M. Heller, D. Kite, and B. Beloff. 1995. "Environmental Accounting Case Study: DuPont." In *Green Ledgers: Case Studies in Corporate Environmental Accounting*, edited by D. Ditz, J. Ranganathan, and D. Banks. Washington, DC: World Resources Institute.

Shonnard, D. R., and D. S. Hiew. 2000. "Comparative Environmental Assessments of VOC Recovery and Recycle Design Alternatives for a Gaseous Waste Stream." *Environmental Science & Technology* 34(24):5222–28.

Shonnard, D. R., A. Lidner, Nhan Nguyen, et al. 2007. "Green Engineering—Integration of Green Chemistry, Pollution Prevention, and Risk-Based Considerations." In *Kent and Reigel's Handbook of Industrial Chemistry and Biotechnology, Volume 1,* 11th ed. Springer Science and Business Media.

UNDSD (United Nations Division of Sustainable Development). 2001. *Indicators of Sustainable Development: Guidelines and Methodologies*. New York: United Nations. www.un.org/esa/sustdev/publications/indisd-mg2001.pdf.

U.S. Congress, Office of Technology Assessment (OTA). 1994. *Industry, Technology and the Environment: Competitive Challenges and Business Opportunities.* OTA-ITE-586. Washington, DC: U.S. Government Printing Office.

U.S. EPA (U.S. Environmental Protection Agency). 1995 *An Introduction to Environmental Accounting as a Business Management Tool: Key Concepts and Terms.* EPA 742-R-95-001. June.

_____. 2011. *Ozone Layer Protection—Science* Web site. www.epa.gov/ozone/science/ods/classone.html and www.epa.gov/ozone/science/ods/classtwo.html. Accessed March 8, 2011.

WCED (World Commission on Environment and Development). 1987. Report of the World Commission on Environment and Development. New York: United Nations.

WHO (World Health Organization). 2000. "Trichloroethylene." Chapter 5.15 in *Air Quality Guidelines, 2nd Edition.* Copenhagen. www.euro.who.int/__data/assets/pdf_file/0005/74732/E71922.pdf.

WMO (World Meteorological Organization). 2007. Scientific Assessment of Ozone Depletion: 2006. Global Ozone Research and Monitoring Project—Report No. 50. Geneva, Switzerland: World Meteorological Organization.

CHAPTER 6

Case Studies

6.1   INTRODUCTION

General engineering design principles that promote sustainability have begun to emerge (Abraham and Nguyen, 2003; Anastas and Zimmerman, 2003; see Chapter 5 of this text). These principles give general design guidance to engineers, but the art and science of designing for sustainability have not yet matured to the point where there is a well-defined set of commonly applied calculation methodologies and mathematical tools.

This current state of development of engineering for sustainability can be compared to the evolution of other fields of engineering. Consider the case of chemical engineering, which emerged late in the 19th century as a field of applied or industrial chemistry (Peppas, 1989). Chemical engineers at the turn of the 20th century studied individual industrial technologies. They learned almost exclusively through case studies. By the middle of the 20th century, however, it became apparent that most chemical processes had common parts, or unit operations. Chemical engineers began to study the design of reactors, distillation columns, and other unit operations, rather than specific processes. By the end of the 20th century sophisticated mathematical tools for modeling chemical reactions, transport phenomena, and thermodynamics had been developed, and these sophisticated design tools were applied at spatial scales from molecular (e.g., modeling the properties of nanomaterials), to the scale of unit operations and chemical processes, to global scales (e.g., modeling global atmospheric chemical reactions). The sophisticated analytical tools available to chemical engineers spanned a wide range of spatial scales, as well as a range of temporal scales, from nanoseconds to decades.

Engineering tools for improving sustainability are a mixture of these three stages. While the field is no longer restricted to just examining case studies, case studies are still revealing new insights. Through the examination of many case studies, common principles have begun to emerge, and they have been the topic of this book. Risk analysis frameworks, life-cycle frameworks, methods for selecting sustainable materials, and a collection of other general principles have gained general acceptance. Some of these principles even have sophisticated analytical tools.

This final chapter of our introduction to engineering for sustainability examines how the principles of Green Engineering can be applied to case studies. Case studies have value in identifying new principles, and in serving as examples of how emergent general principles and analysis tools can be applied. This chapter presents three case studies: (1) biofuels for transportation; (2) transportation, logistics, and supply chains; and (3) sustainable built environments. Questions requiring responses that range from simple calculations to complex, open-ended analyses are embedded in each of the case studies, replacing end-of-chapter problems. Additional case studies that are available in the public domain are listed at the end of the chapter.

6.2   BIOFUELS FOR TRANSPORTATION

Although liquid biofuels for transportation have received increased attention recently, they are not new. A century ago the Ford Model T was designed to run on gasoline, ethanol, and blends of these fuels, becoming a forerunner of modern flexible-fueled vehicles. Rudolf Diesel, the originator of the diesel engine, tested his engine on peanut oil at the 1900 World's Fair in Paris (Bozbas, 2008). The motivation for biofuels a century ago was availability in an era in which petroleum-based fuels were just beginning to be widely available. In contrast, motivations for considering the use of biofuels today include mitigating the emissions of greenhouse gases, rural economic development, domestic jobs, energy security, and balance of trade. In addition, biofuels today include a wider variety of chemical structures, including not only ethanol, but other alcohols, and not only nut oils, but structures derived from organisms such as algae. These new biofuels are possible because of advances in biochemical and thermochemical processing technology, which have also increased the potential sources of biomass feedstocks.

In the United States, federal and state policies are increasingly promoting the use of biofuels. The Energy Independence and Security Act of 2007 (EISA, 2007) set a target of 36 billion gallons per year of biofuel use by 2022. To put this goal in context, in 2010, approximately 140 billion gallons of gasoline and 40 billion gallons per year of diesel were used in the United States. So the goal for biofuel use by 2022 is roughly 20% of the U.S. transportation fuel supply. The European Union is also promoting the use of biofuels, particularly biodiesel (Bozbas, 2008). As these fuels become widely deployed, their sustainability should be carefully considered. This case study will relate the concepts and analysis methods presented in this text to the evaluation of transportation biofuels. The potential environmental benefits and burdens of biofuels derived from a variety of biomass types and through various processing routes will be considered.

6.2.1 The Carbon Cycle and Biofuels

One of the primary motivations for using biofuels is their potential for reducing greenhouse gas (GHG) emissions. Reflecting this, the Renewable Fuel Standard, outlined in the Energy Independence and Security Act (EISA, 2007), defines renewable fuels and advanced biofuels based, in part, on their life-cycle GHG emissions.

Biofuels have the potential to reduce life-cycle GHG emissions because the growth of biomass withdraws carbon dioxide from the atmosphere, through photosynthesis:

$$CO_2 + H_2O + \text{sunlight energy} \rightarrow CHO \text{ (biomass)} + O_2$$

The molecular formula for biomass (CHO) is simplified here by neglecting minor elements such as nitrogen (N), phosphorus (P), and sulfur (S), which constitute DNA, protein, and other cellular components of biomass. In the process of biomass growth, photosynthesis removes CO_2 from the atmosphere, takes up water, fixes carbon and water into solid biomass, and releases oxygen (O_2) to the atmosphere. When biomass dies, organic material in the solid biomass is oxidized either microbially or thermally to form gaseous CO_2 and H_2O (mineralization) with the release of energy, reversing the process of photosynthesis. Figure 6-1 illustrates carbon cycling on a global scale, showing the cycles of photosynthesis and mineralization over terrestrial and marine environments. The carbon reservoirs (atmosphere, ocean, fossil organic carbon, plants, and soil) are shown in gigatons of carbon (GtC), and numbers associated with arrows are estimates of carbon fluxes between reservoirs (GtC/yr). On land, for example, there is a relatively rapid exchange of carbon with the atmosphere that, globally and in the absence of human influences, would be in balance, but this balance may be disrupted locally because of human activity such as deforestation, reforestation, or other land management activity. Coal, natural gas, and petroleum are large stocks of carbon derived from CO_2 sequestered in biomass and other primary producers (plankton) from *tens to hundreds of millions of years ago*. When fossil fuels are combusted, the CO_2 release rate is much greater than the rate at which CO_2 is sequestered again into fossil resources. This imbalance causes an accumulation of CO_2 in the atmosphere from fossil fuel combustion of roughly 3.2 GtC/yr (Figure 6-1). When biofuels produced from biomass are combusted in transportation vehicles, the CO_2 released from biofuels replaces CO_2 *recently* sequestered into biomass. Thus, biofuels have the potential to be *carbon neutral* while providing energy services such as vehicular transportation. However, this view of carbon neutrality of biofuels is overly simplistic in that land management practices and land use change induced by biofuels production may cause additional CO_2 emissions from the land on which the biomass feedstock is grown that would otherwise have not occurred.

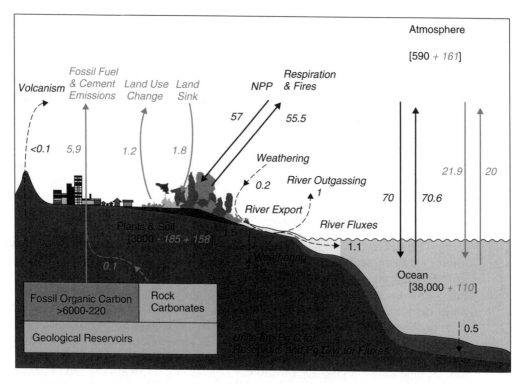

Figure 6-1 The global carbon cycle showing the global carbon reservoirs (atmosphere, ocean, fossil organic carbon, plants, and soil) in petagrams (Pg) of carbon (10^{15} g = 1 GtC = 10^{12} kg) and the annual fluxes and accumulation rates in gigatons of carbon per year, prior to the Industrial Revolution and since the Industrial Revolution (in lighter font). The values shown are approximate, and uncertainties exist as to some of the flow values. (From the National Oceanic and Atmospheric Administration, reported in U.S. Climate Change Science Program, 2007); NPP = net primary production (photosynthesis))

Question 1

If carbon accumulates in the atmosphere at 3.2 GtC/yr, how long will it be until the atmosphere contains 1200 GtC (approximately double the preindustrial level)? Repeat the calculation assuming that the amount of carbon entering the atmosphere due to fossil fuel combustion increases by 2% per year from its present level of 5.9 GtC/yr (assume that all other flows remain constant).

In the most recent report from the IPCC, CO_2 was listed as the most important anthropogenic greenhouse gas, and fossil fuel combustion was identified as the most common anthropogenic source (IPCC, 2007). The concentration of CO_2 has increased from a preindustrial value of 280 ppm to 390 ppm in 2010 (NOAA, 2010), the highest level in the last 650,000 years, and average global temperature increased

0.76°C (from ~13.75°C to 14.5°C) over the same time period (IPCC, 2007). The IPCC projects that global warming will continue to increase at a rate of +0.2°C per decade for the next 20 years and then, even if all man-made GHG emissions are eliminated, warming will continue for centuries because of the long timescales of climate processes.

6.2.2   Feedstocks for Biofuels

Currently, most biofuels used as replacements for gasoline are produced by fermentation of cornstarch, or sugar extracted from cane, producing ethanol as the gasoline replacement. Biodiesel is produced from various plant oils. However, recent technological advancements have made possible a much wider array of biomass feedstocks for production of liquid transportation fuels from biomass. In 2005, a U.S. government study estimated that 1 billion tons of biomass, primarily biomass that is currently viewed as waste, is available on a sustainable annual basis in the United States (Perlack et al., 2005).

Question 2

> If roughly 50% of the billion tons of biomass available in the United States is carbon, and if roughly 50% of that carbon can be converted to liquid fuels (typical of gasification followed by Fischer-Tropsch synthesis processes), estimate the fraction of the 180 billion gallons of gasoline and diesel used in the United States that could come from waste biomass. Calculate the yield of fuel per ton of biomass.

Figure 6-2 shows estimates of annual production of biomass from forest and agricultural lands. Of all anticipated feedstocks, over 90% is lignocellulosic (woody) biomass as opposed to crop grains such as corn, sugar, and soybeans. Biomass from forest lands ranges from low-productivity logging residues (0.25 to 0.50 dry tons/ acre/yr) to high-productivity energy crops such as hybrid poplar and willow (up to 10 dry tons/acre/yr). On agricultural lands, productivities range from lower-productivity crop residues such as corn stover (2 dry tons/acre/yr) to higher-productivity perennial energy crops such as switchgrass and miscanthus (10 dry tons/acre/yr).

The estimates in Figure 6-2 are to be interpreted with caution as the values are affected by study assumptions, some of which include projected increases in crop productivity per acre, modifications to crop cultivation practices, fertilizer application rates, and crop residue collection efficiency. The study also assumed that all of the identified biomass would actually be available for conversion to biofuels, but availability is very sensitive to local economic conditions and landowner choices for the use of these resources. Feedstock estimates should be interpreted as biomass production potential as opposed to actual availability. In addition to these study limitations, other forms of biomass were not included in the feedstocks considered. Algae is gaining interest as a highly productive source of bio-oil, which can be grown in contained ponds using seawater or water from deep saline aquifers.

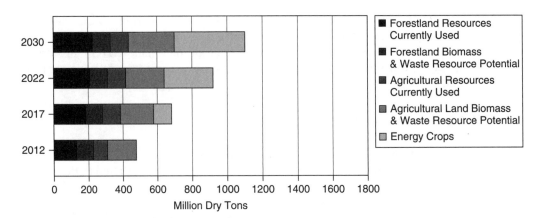

Figure 6-2 A summary of currently used and potential resources at $60 per dry ton or less (U.S. Department of Energy, 2011b)

Per-acre productivities of algae can be very high, from 10 to 100 times greater than terrestrial energy crop productivities (Greenwell et al., 2010). Other energy crops were not covered in the "billion-ton study update" (U.S. DOE, 2011b) report such as jatropha, camelina, and castor, which are suitable for cultivation on either marginal lands or as rotation crops among food crops such as winter wheat (Shonnard et al., 2010) and require lower inputs of water and fertilizer than conventional crops. However, in order to take advantage of all of these biomass feedstocks, new technology must be developed to convert the more recalcitrant woody and plant oil biomass into liquid transportation fuels.

Question 3

One of the challenges of generating fuels from biomass is that the sources of biomass must be transported to centralized facilities for processing. Both trucks and trains are used to transport the material. Calculate the fraction of the fuel yield per ton of biomass (Question 2) that is expended in aggregating the biomass in a single facility. For one scenario, assume that the average transport distance is 100 km (by truck). For a second scenario, assume that the average transport distance is 50 km by truck and 500 km by train. Assume that fuel consumption for truck and train transport is 0.027 and 0.0065 liters of fuel per ton of biomass transported per kilometer, respectively (NREL, 2011).

6.2.3   Processing Routes for Biomass to Biofuels

Processing routes for conversion of biomass to biofuels have traditionally been organized into two categories, depending on the agents for transformation and reaction conditions: biochemical and thermochemical. Biochemical conversion processes employ biological catalysts such as enzymes at mild temperatures to produce sugars from the original biomass and then ferment the sugars into oxygenated biofuels using microorganisms (Houghton et al., 2006). The choice of microorganism should take into account the types of structures in the original bio-mass, the sugars fermented, and the specific fermentation products produced. Thermochemical conversion processes employ chemical catalysts and are, for the most part, carried out at higher temperatures and pressures (NSF, 2008). These high-temperature reactions exhibit much shorter reaction times than biochemical conversion processes but often yield a wide variety of products rather than specific chemicals, such as ethanol, that can result from fermentation.

Figure 6-3 is an overview diagram showing the main conversion steps, feed-stocks, intermediates, and products for conventional biofuels (ethanol from starch crops and cane, biodiesel from triglycerides in soybeans) and for advanced biofuels. The dry mill corn ethanol process produces an intermediate glucose product and a final product of ethanol plus dry distiller grain solids (DDGS), which are marketed as animal feed. The overall fermentation reaction is given by $C_6H_{12}O_6 \rightarrow 2\ C_2H_5OH + 2\ CO_2$. Two carbons from the sugar molecule are lost as carbon dioxide, but the energy content of the two ethanol molecules is substantially higher than that of the sugar feedstock. Process energy for corn ethanol production is typically from natural gas for steam production, and electricity is from the local grid. Because of the use of these process energy resources, corn ethanol has a relatively large fossil energy demand (ratio of fossil energy required for all processing steps per unit of energy in ethanol produced) of approximately 0.4 (Shapouri et al., 2010) or higher. The production rate of corn grain ethanol in the United States in 2009 was approximately 12 billion gal/yr (RFA, 2010a), and from sugarcane in Brazil (in 2008) was about 6.6 billion gal/yr (RFA, 2010b).

Biodiesel is a methyl ester of fatty acids derived from plant oils. The biodiesel reaction can be simply described as triglyceride + methanol $\rightarrow$ 3 fatty acid methyl esters + glycerol ($CH_2OH\text{-}CHOH\text{-}CH_2OH$). Methanol is almost exclusively pro-duced from natural gas (fossil origin), and co-products of biodiesel production include glycerol and a residue from the oil extraction step (soymeal, for example), which is often marketed as an animal feed. The key intermediate is a plant oil obtained from the oil extraction step.

Advanced biofuels are produced biochemically or through thermochemical conversion processing, but there are large differences from conventional biofuels processes. Hydrolysis of woody biomass is more difficult than for crop starch and will yield a mixture of 5- and 6-carbon sugars (Houghton et al., 2006), yet not many naturally occurring microorganisms are able to readily ferment 5-carbon sugars. Recent advances in metabolic engineering and systems biology of microorganisms have created unique metabolic pathways within microorganisms so that mixtures of sugars obtained from lignocellulosic biomass can be fermented into oxygenated biofuels, such as ethanol and butanol, and also into hydrocarbon fuels (Steen et al., 2010). In 2010, over 20 cellulosic ethanol commercial or demonstration projects were in development or construction with maximum production capacity of greater than 400 million gallons of cellulosic ethanol per year (RFA, 2010c). Thermochemical processing of triglycerides from plants and algae consists of mechanical and solvent extraction to recover crude plant bio-oil followed by catalytic upgrading through hydrotreatment to hydrocarbon biofuels: green gasoline, green diesel, and renewable jet. Although these hydrocarbon fuels are similar to petroleum fuels, they often have superior materials compatibility, stability, combustion, and emission properties compared to conventional biofuels and to fossil fuels (Kalnes et al., 2007). Thermochemical conversion of woody (lignocellulosic) biomass starts with gasification or pyrolysis, both being high-temperature and—pressure thermal decomposition processes. Intermediate synthesis gas (gasification) and pyrolysis bio-oil (Py-Oil/ HydroPy-Oil) are upgraded catalytically in the presence of hydrogen to hydrocarbon biofuels.

There are advantages and limitations for each conventional and advanced biofuel conversion route, some being more technologically feasible and economical and others possibly being more environmentally benign (lower GHG emissions over the life cycle). Productivities per amount of biomass feedstock can also vary widely. Ultimately, the success of biofuels will depend on many factors, including social acceptability, compatibility with existing vehicular and processing infrastructure, economics, and environmental performance. The next section describes recent comparisons of environmental performance for conventional fossil fuels and future advanced fuels: biofuels and fossil-based.

Question 4 (open-ended)

Biofuels must have a variety of properties in order to serve as drop-in replacements for conventional fuels. Identify substitutes for gasoline that have equal or higher octane numbers compared to iso-octane (octane numbers are the quality indicator used in the sale of gasoline; estimation methods for octane number are given by Ghosh et al., 2006), equal or lower vapor pressure compared to iso-octane (a typical gasoline component), equal or greater biodegradability compared to iso-octane, equal or lower water solubility than iso-octane, and molecular weights lower than 170. Most of these properties can be estimated using the methods described by Allen and Shonnard (2001) or by the EPI Suite software (available at www.epa.gov/oppt/greenengineering). The molecule should contain only carbon, hydrogen, and oxygen, and when added to gasoline at the 15% level, it should not increase aromatic content by more than 5%, since there is a cap on the fraction of aromatic species in gasoline. Subject to these constraints, maximize the octane number.

Question 5 (open-ended)

Review the recent scientific literature on biofuels and identify a biofuel production pathway. Write a one-page summary, describing the feedstock, the approximate amount of the feedstock available, the conversion route, and the types of fuels and by-products produced.

Conventional Biofuels

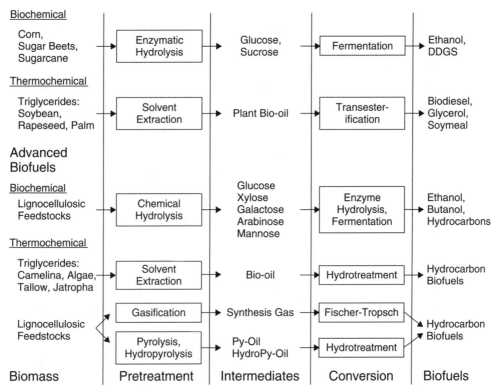

Figure 6-3   Conventional and advanced process technologies for biofuels production

6.2.4   Biofuel Life Cycles

Life-cycle assessment (LCA) is a tool that is particularly well suited to the characterization of the environmental impacts of biofuels. LCA has been used to guide research and development toward more productive synthesis routes and will continue to be used to refine approaches for biomass feedstock cultivation, harvesting, and transportation, as well as for production and use of biofuel products. This section presents a life-cycle comparison of conventional and advanced fuels, both fossil- and bio-based (Koers et al., 2009). Additional analyses are available at www. utexas.edu/research/cem/projects/epa_report.html.

Figure 6-4 shows key stages in the production of biodiesel, green diesel, petroleum diesel, and synthesis diesel. Important inputs, products, and co-products are shown at various life-cycle stages. A functional unit was chosen as 1 MJ of energy in the fuel product in order to develop life-cycle inventory data. Inventory data for each product system were obtained from reports and other sources, as described in

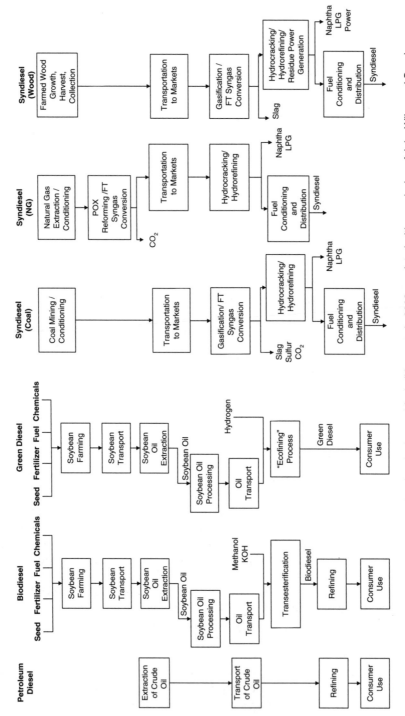

Figure 6-4 Life-cycle diagrams for conventional and advanced fuels (Koers et al., 2009; reprinted with permission of John Wiley and Sons)

Koers et al. (2009). Key study assumptions and inventory data sources are shown in Table 6-1. Scope for each fuel product is "cradle to grave," from extraction from nature to product combustion in engines. One study (Hill et al., 2006) stands out from the others by including effects beyond the product systems shown in Figure 6-4, including the impacts of fuel use to drive to and from the farm and impacts of manufacture of farm equipment and buildings. Impacts contributing to global warming were assessed using global warming potentials (GWPs) of greenhouse gases relative to CO_2 (GWPs: $CO_2 = 1$, $CH_4 = 21$, $N_2O = 300$, refrigerants = 1000s).

A comparison of GHG emissions expressed in CO_2 equivalents for the conventional and advanced fuels is shown in Figures 6-5 and 6-6. Based on this assessment, green diesel is comparable to or lower in GHG emissions than biodiesel, and results from the DOE study (Sheehan et al., 1998) data are much lower, suggesting that neglecting N_2O emissions from soybean production introduces a large error. Impacts of petroleum diesel are for the most part over 50% higher than those of green diesel and biodiesel. Synthesis diesel from coal is estimated to have over twice the impact on GHG emissions as petroleum diesel, synthesis diesel from natural gas is about 20% greater than petroleum diesel, whereas synthesis diesel from wood emits approximately 10% of the GHG emissions compared to petroleum diesel. An analysis such as this should include sensitivity studies, identifying key variables and assumptions and using alternative values for these parameters.

Question 6

A tool for performing sensitivity analyses is available from Argonne National Laboratory, the GREET model. GREET, the Greenhouse Gases, Regulated Emissions, and Energy Use in Transportation Model, can be downloaded from http://greet.es.anl.gov/.

a. Use the GREET model to estimate GHG emissions for a mix of the following fuels (use default settings for all fuels):
    i.   100% ethanol from cornstarch
    ii. 100% ethanol from woody biomass
    iii. 100% ethanol from forest residue

Table 6-1   Key Study Assumptions and Inventory Data Sources

Fuel[a]	Feedstock	N_2O Included?	Data Source
GD, BD (DOE)	Soybean	No	Sheehan et al., 1998
GD, BD (CONCAWE)	Rapeseed	Yes	CONCAWE, 2007
GD, BD (PNAS)	Soybean	Yes	Hill et al. 2006
Syndiesel (CONCAWE)	Coal	Yes	CONCAWE, 2007
Syndiesel (CONCAWE)	NG	Yes	CONCAWE, 2007
Syndiesel (CONCAWE)	Wood	Yes	CONCAWE, 2007
Petroleum diesel	Crude oil	Yes	SimaPro software

[a]GD: green diesel; BD: biodiesel; Syndiesel: synthesis diesel produced by gasification and Fischer-Tropsch conversion to hydrocarbon biofuels

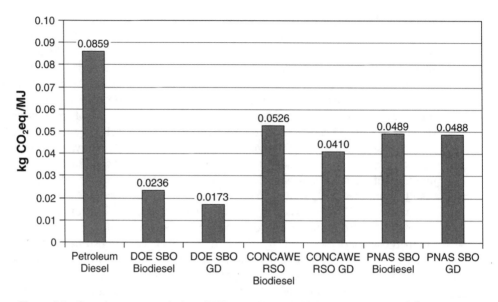

Figure 6-5 Greenhouse gas emissions (SBO = soybean oil, RSO = rapeseed oil, GD = green diesel) (Koers et al., 2009; reprinted with permission of John Wiley and Sons)

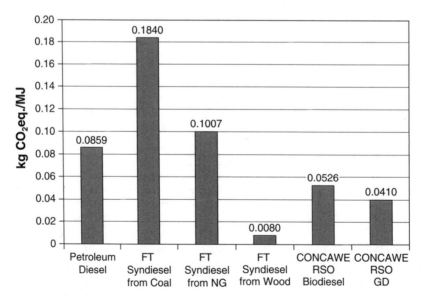

Figure 6-6 Greenhouse gas emissions (NG = natural gas, GD = green diesel). Displacement allocation used for all substitute diesel fuels. (Koers et al., 2009; reprinted with permission of John Wiley and Sons)

Express your results as grams of the various greenhouse gases per million BTU of lower heating value of the fuel.

b. For 100% ethanol from woody biomass, evaluate the sensitivity of GHG emissions to the choice of electricity mix (U.S. average, northeastern United States, or California)

c. Examine the impact of including, versus not including, land use impact.

6.2.5 Cautionary Tales and Biofuels

The results presented in Section 6.2.4 show GHG emission advantages for biodiesel, green diesel, and synthesis diesel from wood compared to conventional petroleum diesel and synthesis diesel from fossil resources. In these analyses, biomass carbon was considered as climate neutral, and therefore only fossil-derived CO_2 emissions were accounted for. However, there is a growing concern that not only fossil-derived but also biomass-derived CO_2 emissions must be included in life-cycle assessments of bio-based products, and that CO_2 emissions from changes in land management are important (Searchinger et al., 2009). Furthermore, the use of fertilizers for biomass energy crop cultivation can contribute to nutrient loadings of streams, lakes, and rivers. Finally, the use of water in biomass production can be significant. The concluding sections of this case study will introduce these consequences of large-scale biofuel production.

Land Use Change: Direct and Indirect Effects

Land use change has occurred often throughout history, most often when natural landscapes are cleared of native vegetation and come under agricultural cultivation. Today, deforestation of tropical forests is the most important example of this (Figure 6-1), resulting in a significant flux of man-made CO_2 into the atmosphere (1 GtC/yr). As biofuels production increases significantly, there is great potential for natural landscapes to come under energy crop cultivation. When this occurs, often the amount of biomass on the land after conversion to energy crops can be much less than before. Figure 6-7 shows the evolution of cultivation on land in the Brazilian rainforest between 1975 and 1992. During land conversion, vegetation is often cleared and biomass burned, releasing a large flux of above- and below-ground carbon as CO_2 to the atmosphere. The conversion of land into fuel crop production is referred to as a *direct land use change*. If agricultural lands are converted from food production to fuel crop production, an *indirect land use change* occurs. Indirect land use change emissions of CO_2 come about as a consequence of large-scale biofuel production on lands previously used for agriculture because demand for food is inelastic (i.e., must always be met). Any lands previously used for food production that are converted to non-food uses (such as energy cropping for biofuel) will cause land use change elsewhere in the world from a natural state to agricultural cultivation. When this conversion occurs, often a large flux of CO_2 from land clearing will result. Predicting these future land conversions is very

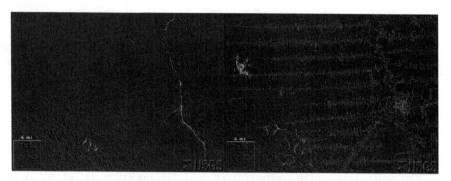

Figure 6-7 Satellite images of Rondonia, in Brazil, showing the increase in cultivation between 1975 (left) and 1992 (right) (USGS, 2011c)

complicated, requiring an understanding of market forces, economic drivers, and societal behavior in countries around the world. A modeling approach adopted by the U.S. EPA in a final rule-making document for the federal Renewable Fuel Standards (U.S. EPA, 2010) estimates that indirect land use change emissions will equal 0.030 kg CO_2 eq./MJ corn ethanol, 0.041 kg CO_2 eq./MJ soybean biodiesel, and 0.014 kg CO_2 eq./MJ switchgrass ethanol.

Comparing these values to the results reported in Figures 6-5 and 6-6 demonstrates that indirect land use change emissions can be a significant, and at times a dominant, contribution to a biofuel's GHG emissions.

Question 7

The EPA performed a Regulatory Impact Assessment (RIA) for the Renewable Fuel Standard in the Energy Independence and Security Act. In the RIA, the EPA estimated GHG emissions for a variety of fuels (U.S. EPA, 2010). Compare the estimates for corn ethanol to those for sugarcane and ethanol from switchgrass.

Eutrophication from Nutrient Runoff and Leaching

Growing biomass requires macronutrients such as nitrogen, phosphorus, and potassium (NPK) to satisfy elemental requirements of biomass components (DNA, proteins, ATP, etc.). These nutrients are added to fields during crop cultivation. Some energy crops are heavily reliant on fertilizers, pesticides, and soil conditioners for high-productivity cultivation (corn), whereas other feedstocks are provided without external inputs (logging residues). There is a broad range of required nutrient inputs depending on biomass type, soil quality, and other factors. A portion of the fertilizers added to fields will leach into groundwater or run off fields into streams, rivers, and lakes. Depending on fertilizer type, climate, and soil conditions, as much as 50% of applied nitrogen fertilizer may run off from fields into streams or leach into groundwater. An emission factor of 30% for N fertilizers is recommended by the Intergovernmental Panel on Climate Change (IPCC, 2006) in one method. These pollutants stimulate the growth of naturally occurring algae, which degrades the clarity of water and causes oxygen depletion when the algae die and are mineralized by microbial action. These nutrients also exert negative impacts at the mouths of major rivers on all continents of the world, creating zones of oxygen depletion termed *dead zones*. Figure 6-8 shows the Mississippi River basin feeding fertilizer runoff to the Gulf of Mexico and depressed readings of oxygen content in ocean water. Solutions to dead zones include building wetlands and buffer zones to capture runoff and to create slow-release fertilizer formulations that deliver nutrients to plants more accurately. High costs impede the adoption of these and other responses to eutrophication.

Water Use

Growing biofuels requires water. In some regions water is provided by rainfall, but as increasingly marginal lands are used to satisfy both food and fuel crop needs, it is likely that increasing amounts of irrigation would be required. Historically, data on water use have been difficult to obtain; however, as water becomes an increasingly scarce resource, accounting for life-cycle water use will become increasingly important.

When water is accounted for, a distinction is generally made between withdrawals and consumption. For growing biofuel crops, withdrawal is a result of irrigation, while consumption includes evapo-transpiration and the conversion of water into biomass. Table 6-2 shows estimates of water use for a variety of crops, separated into withdrawal and consumption. These estimates indicate that water use for biofuel crops can amount to thousands of gallons per gallon of fuel produced. In comparison, the extraction of crude oil has a relatively small water footprint,

Table 6-2 Water Use for Different Feedstocks during Crop Growing

Feedstock	Freshwater (10^6 L/ha-yr)		Fuel Yield	Fuel Yield (Petroleum Equivalent)	Volume of Freshwater/ Volume of Fuel (Petroleum Equivalent)	
	Withdrawn	Consumed	L/ha-yr	L/ha-yr	Consumed	Withdrawn
Corn for grain	0.45	7	3787	2490	2811	181
Sugarcane for sugar	3.59	6.7	6510	4281	1565	839
Soy for beans	0.18	6	599	558	10,759	323
Crude oil					3	8

Source: Murphy, 2011

MISSISSIPPI RIVER BASIN

ZONE DEFENSE

Since much of the problem can be traced back to farm runoff, efforts to control the dead zone often target fertilizer usage in agricultural areas throughout the Mississippi River basin.

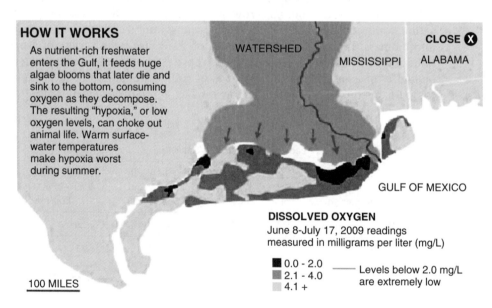

HOW IT WORKS

As nutrient-rich freshwater enters the Gulf, it feeds huge algae blooms that later die and sink to the bottom, consuming oxygen as they decompose. The resulting "hypoxia," or low oxygen levels, can choke out animal life. Warm surface-water temperatures make hypoxia worst during summer.

DISSOLVED OXYGEN
June 8-July 17, 2009 readings measured in milligrams per liter (mg/L)

- 0.0 - 2.0
- 2.1 - 4.0
- 4.1 +

—— Levels below 2.0 mg/L are extremely low

100 MILES

Figure 6-8   Dead zone formation can be a result of fertilizer runoff. (http://ecowatch.ncddc.noaa.gov/hypoxia)

although there is a great deal of uncertainty regarding the amount of water used in the production of crude oil.

Question 8

Based on the data in Table 6-2, estimate the gallons of water, per mile of travel, required for a vehicle that has a fuel efficiency of 30 mpg of fuel. Perform the calculation for corn, sugarcane, soy oil, and crude oil.

6.2.6 Summary of Sustainability of Biofuels

Biomass is a potential feedstock that is created by photosynthesis, in a process that removes CO_2 from the atmosphere. Combustion of biofuels from biomass returns carbon to the atmosphere in a closed cycle. As much as 1 billion dry tons of biomass are available per year in the United States, a quantity that could replace a significant fraction of current petroleum use (2005 data), and most of this biomass is of the woody type. Conventional and advanced biofuels processing technologies are under development with the goal of overcoming technical and economic barriers to large-scale commercial biofuels production. Studies of the environmental benefits of conventional and advanced biofuels are ongoing and have encouraging features, but challenges exist on managing biomass production to ensure that CO_2 emissions from land use change are not severe, that water consumption can be accommodated, and that water quality is protected from excessive nutrient runoff from energy crop production.

6.3   TRANSPORTATION, LOGISTICS, AND SUPPLY CHAINS

Logistics is the management of the flow of goods and services within a production and distribution system between points of origin and points of use. Logistics has origins in military applications, stemming from the need to supply troops on the move with ammunition, food, and other items. Outside of military uses, logistics figures prominently in manufacturing, distribution of produced goods to retail outlets, and managing business projects. In manufacturing logistics, the focus of effort is to ensure that the manufacturing and assembly systems are supplied with the right components at the right time and in the right order. In retail logistics, supply chain choices involve levels of inventory, transportation, locations of warehouses, tracking of goods, and management. Logistics is a part of all engineering disciplines but is most important in the fields of industrial engineering, mechanical engineering, and transportation studies.

The case studies in this section explore sustainability issues in transportation logistics. Because these choices involve supply chains, life-cycle thinking is implicit in the design decisions.

6.3.1 A Limited Life-Cycle Assessment of Garment Design, Manufacture, and Distribution

The first case study in logistics that will be considered in this section (Hopkins et al., 1994) examines the transport of the materials for a fleece pullover and final garments along their supply chains. Studies of garment life cycles have generally not found transportation to be a significant factor because road and rail transport are often used as the primary transport modes (see Table 6-3). For example, it was found that a 0.123 lb polyester woman's blouse consumed 10,528 BTU for the manufacturing step (Franklin Associates, 1993) and that transportation contributed only about 5% to this total. The importance of transportation can change significantly, however, if air transport is used to achieve just-in-time deliveries. This suggests that optimum logistics will involve balancing the impacts of rapid transportation with the impacts of warehousing and storage of inventory along the supply chain.

The first step in this case study will be to examine the impacts of transportation of the garments. The garment weight is 1.28 lb and the energy required for garment manufacture is assumed to be approximately 100,000 BTU.

6.3.2 Alternatives for Garment Transport Logistics

Two scenarios will be evaluated for the transport of fabric to the location of garment assembly, and then transport of the finished garment to warehousing. In scenario 1, fabric is transported from a mill in Massachusetts to a sewing factory in San Francisco by truck (road), and the finished garment is then shipped to Ventura, California, again by road transport. In scenario 2, fabric is transported by rail from Massachusetts to Miami, and then by ship to the location of sewing, Jamaica. Finished garments are transported from Jamaica to Miami by air freight, and then to Ventura by road. Table 6-4 shows transportation distances and modes for fabric and garment transport.

Question 9

Use the energy consumption data in Table 6-3 to estimate the energy used to transport the material for one garment along the supply chains described in Table 6-4. Note that

Table 6-3 Energy Use by Transportation Mode

Mode of Transport	BTU/ton-mile
Ocean freighter	396
Rail freight	411
Road freight	3357
Air freight (all cargo)	21,670

Source: Davis and Strong, 1993

Table 6-4 Data for Fabric and Garment Transport from Massachusetts to Ventura via San Francisco or Jamaica

				Shipping Distances		
Item	Weight	Route/ Destination	Road Miles	Ship Miles	Air Miles	Rail Miles
One roll[a] of fabric from mill	57 lbs	MA to San Francisco	3000			
One roll[a] of fabric from mill	57 lbs	MA to Jamaica		600 (Miami to Jamaica)		1200 (MA to Miami)
One box of 18 garments	28.6 lbs	San Francisco to Ventura	400			
One box of 18 garments	28.6 lbs	Jamaica to Ventura via Miami	3000 (Miami to Ventura)		587 (Jamaica to Miami)	

[a]One roll of fabric produces 30 garments.

the energy consumption factors used in Table 6-3 are actual liquid fuel use during transport and do not include energy consumed to extract, transport, and produce the liquid fuel, the so-called upstream processing steps.

Table 6-4 maps the transportation required to move a garment to a product warehouse, but the garment still needs to be transported to the consumer. Many products are sold to customers through retail outlets, and there are multiple modes of transportation that could be used. One pathway is to ship garments to retail outlets in cardboard boxes with 18 to 20 garments per box. Garments are also shipped directly to the final customer through mail order, and there are several shipping pathways, each with unique cost and time-of-delivery features. Two scenarios for the transport of a garment from the headquarters in Ventura, California, to the final customer are described in Table 6-5: to a retail outlet in Boston, Massachusetts, or directly to a customer through mail order to Denver, Colorado.

Question 10

Use the energy consumption data in Table 6-3 to estimate the energy used for transporting one garment from the warehouse to a customer along the supply chains described in Table 6-4. Also estimate the cost of each pathway and compare the transportation energy requirements to the estimated energy requirements for garment manufacture (100,000 BTU).

Table 6-5 Data for Garment Transport to Retail Outlet or Direct to Customer

Shipment: One garment (man's large)
Weight: 1.4 lbs (including packaging)
Origin: Ventura, CA

Scenario	Service	Route	Mode of Transport	Distance (miles)	Cost ($)	Transit Time (days)
Retail outlet	Ground	Ventura to Grande Vista, CA to Boston, via Chicago	Truck	100	$1/garment	9–10
			Train	3244		
	Second-day air	Ventura to Ontario, CA to Philadel-phia to Boston	Truck	100	$2/garment	2
			Air freight	2600		
	Next-day air	Ventura to Burbank to Louis-ville to Boston	Truck	60	$3/garment	1
			Air freight	2700		
Mail-order customer	Ground	Ventura to Grande Vista, CA to Denver	Truck	100	$3/garment	3
			Train	1353		
	Second-day air	Ventura to Ontario, CA to Denver	Truck	100	$8/garment	2
			Air freight	800		
	Next-day air	Ventura to Ontario, CA to Denver	Truck	100	$18/garment	1
			Air freight	800		

Question 11

Assume that warehousing consumes 4% of the manufacturing energy of the garment. How do the differences in energy use in the manufacturing transportation scenarios compare to the total warehousing energy demand? Describe how you might approach an integrated model of energy consumption in transportation and warehousing.

6.3.3 Life Cycles, Materials Use, and Transportation Logistics of Running Shoes

A second logistics case study (Morris, 2011) examines global transport of shoe components and the controversial use of a material in the sole of running shoes. This case will examine GHG emissions over the life cycle of the shoes, comparing the impacts of transport to the impacts of the selection of one critical material.

The system boundaries include the extraction of virgin materials, the transportation of the materials to their place of assembly, the manufacture of the shoes, and the shipping of finished shoes to the United States. Data for the case study were derived from sourcemap.com,[1] an open-access site devoted to tracing distances and materials of products. Data were also available on GreenXchange,[2] the Web site dedicated to sharing green design innovations pioneered in 2008 by shoe manufacturers.

Question 12

Calculate the fuel use associated with the transport of 2 million pairs of shoes. Assume that air, ship, truck, and train are all used in the transport over the life cycle. Further assume that the amounts of air, ship, truck, and train transport used per pair of shoes are 4.3, 17, 1.0, and 2.7 tkm, respectively (recall that a ton-kilometer is the transport of 1 kg of material 1000 km or the transport of 1000 kg of material 1 km). Fuel usages for air, ship, truck, and train transport are 0.42, 0.0049, 0.027, and 0.0065 liters of diesel/jet fuel per ton of shoes transported per kilometer, respectively (NREL, 2011).

Question 13

Based on the data in Question 12, provide a rough estimate of the distance that the shoe components traveled.

Question 14

Calculate the GHG emissions associated with the transport of 2 million pairs of shoes. Assume that the life-cycle GHG emissions associated with diesel/jet fuel are 95 kg CO_2 equivalents per million BTU of heating value of fuel (Skone and Gerdes, 2008) and that the heating value of the fuel is 35,000 BTU/L.

Question 15

Assume that the initial design of the shoes incorporated 0.00073 kg of sulfur hexafluoride per pair in the form of gas inclusions in the soles. Assume that all of this sulfur

---

1. www.sourcemap.com.
2. http.//greenxchange.force.com/vGXsearch?keywords=&category=&company=001A00000025TTWIA4.

hexafluoride escapes to the atmosphere when the shoes reach the end of their life. Calculate the GHG impacts (in CO_2 equivalents per 2 million pairs) of using this material and compare the estimate to the total GHG emissions due to transport. Describe methods for reducing the GHG impacts of the shoe.

6.3.4 Sustainability and Logistics

In these transportation logistics case studies, the analyses and impacts focused on environmental issues as opposed to broader sustainability issues, such as economic factors and societal concerns. Nevertheless, the cases demonstrated that the extent and mode of transportation used in supply chains can be very important in determining a product's footprint. The use of global supply chains will have implications for environmental performance, costs, and employment and social structures around the world. Although it was not possible to perform a full sustainability analysis for these case studies, it is becoming clear that engineers of all disciplines will become increasingly involved in making decisions about supply chains, and those decisions will have global environmental, economic, and societal impacts.

6.4   SUSTAINABLE BUILT ENVIRONMENTS

Buildings are familiar and important elements of modern life, yet the impacts of buildings on the environment and on human health and well-being are not well understood by most building users and are often not reflected in society's approaches to protecting the environment and human health. For example, many environmental regulations are based on outdoor conditions, yet most citizens of the United States and other developed countries spend more than 90% of their time in buildings (U.S. EPA, 1987). While environmental conditions in buildings are to a small degree dependent on pollutants that are brought in from outside, many pollutants are found at greater concentrations indoors than outdoors, because of emissions released from materials within the building (e.g., walls, carpets, furniture) or activities done in the building (e.g., cooking, smoking).

In addition to buildings being an important living environment, their construction and operation are directly and indirectly responsible for the consumption of significant fractions of national energy and material flows. This consumption has environmental, economic, and social impacts.

The case study presented in this section explores the importance of buildings for the use of energy resources and materials and how a variety of sustainability criteria can be used to evaluate building performance. The use of U.S. statistics will dominate the presentation, but the interpretations derived from the U.S. data are expected to be relevant to other developed and developing countries. This case study will also introduce methods and tools that can be used to evaluate the sustainability of building designs and will briefly explore how trade-offs are often encountered (e.g., improving indoor air quality versus reducing energy losses by recirculating indoor air) in building design decisions.

6.4.1 Energy Consumed for Building Operation

Residential and commercial buildings are important end users of energy in the United States, accounting for approximately 20% of end-use national consumption (see Figure 1-3). Table 6-6 shows energy use data for residential and commercial buildings, by application and by energy type. Most energy used in buildings is in the form of natural gas and electricity, with minor contributions from biomass and petroleum. Because a substantial fraction of the energy use in buildings comes from electricity, a more comprehensive view of building energy use would include the energy losses associated with generating electricity and producing and delivering natural gas. If these energy losses are accounted for, then buildings in the United States consume about 40% of total primary energy (94.6 quadrillion BTU; U.S. DOE EIA, 2010). This includes 72% of U.S. electricity consumption and 39% of annual U.S. CO_2 emissions.

Space heating, lighting, and space cooling are the largest building energy consumers, constituting together about 37% of total building energy use. Of less importance are water heating, refrigeration, electronics, and ventilation. Ventilation, although the smallest category, contributes indirectly to larger energy use categories such as space heating and cooling by exporting building air to the environment and importing outside air into the building, both of which require inputs of primary energy for heating and cooling, depending on the season.

Knowing the major energy uses in buildings can inform the energy-efficient design of buildings. For example, if a building is in close proximity to an industrial facility, sharing industrial waste heat could offset fossil energy resources for building heating. Energy for building heating can also be saved through greater use of insulation materials in walls and windows. Energy impacts of lighting can be reduced through the use of natural light.

Question 16

Review the energy use data for operation of buildings in Table 6-6 and propose energy savings or pollution prevention approaches. An Internet search on "energy savings for buildings" will lead to multiple Web sites (e.g., www1.eere.energy.gov/buildings) with suggested energy savings that are no-cost, low-cost, or higher-cost solutions. Sophisticated building energy modeling tools are also available, including public domain models such as EnergyPlus from the U.S. Department of Energy's Energy Efficiency and Renewable Energy program (http://apps1.eere.energy.gov/buildings/energyplus) and as described later in this case study.

Table 6-6 2010 U.S. Buildings Energy End-Use Splits, by Fuel Type (Quadrillion BTU)

	Natural Gas	Fuel Oil[a]	LPG	Other Fuel[b]	Renw. En.[c]	Site Electric	Site Total	Site Percent	Primary Electric[d]	Primary Total	Primary Percent
Space heating[e]	4.90	0.68	0.26	0.09	0.54	0.60	7.07	35.4%	1.89	8.36	20.7%
Lighting						1.69	1.69	8.5%	5.42	5.42	13.4%
Space cooling	0.04					0.51	0.55	2.8%	5.31	5.35	13.2%
Water heating	1.80	0.12	0.08		0.03	0.53	2.55	12.8%	1.67	3.69	9.1%
Refrigeration[f]						0.83	0.83	4.2%	2.61	2.61	6.5%
Electronics[g]						1.73	1.73	8.6%	1.89	1.89	4.7%
Ventilation[h]						0.13	0.13	0.7%	1.60	1.60	4.0%
Computers						0.39	0.39	2.0%	1.22	1.22	3.0%
Wet cleaning[i]	0.05					0.60	0.65	3.3%	0.97	1.02	2.5%
Cooking	0.40		0.03			0.31	0.74	3.7%	0.41	0.83	2.1%
Other[j]	0.30	0.01	0.30	0.04	0.01	1.55	2.22	11.1%	4.85	5.52	13.7%
Adjust to SEDS[k]	0.58	0.15				0.69	1.42	7.1%	2.15	2.89	7.2%
Total	**8.08**	**0.96**	**0.67**	**0.14**	**0.58**	**9.57**	**20.00**	**100%**	**29.98**	**40.40**	**100%**

[a] Includes distillate fuel oil (0.92 quad) and residual fuel oil (0.04 quad).

[b] Kerosene (0.03 quad) and coal (0.07 quad) are assumed attributable to space heating. Motor gasoline (0.04 quad) assumed attributable to other end uses.

[c] Composed of wood space heating (0.42 quad), biomass (0.11 quad), solar water heating (0.03 quad), geothermal space heating (0.01 quad), solar photovoltaics (PV) (0.01 quad), and wind (less than 0.01 quad).

[d] Site-to-source electricity conversion (due to generation and transmission losses) = 3.13.

[e] Includes furnace fans (0.14 quad).

[f] Includes refrigerators (2.37 quads) and freezers (0.25 quad). Includes commercial refrigeration.

[g] Includes color television (1.07 quads).

[h] Commercial only; residential fan and pump energy use included proportionately in space heating and cooling.

[i] Includes clothes washers (0.10 quad), natural gas clothes dryers (0.05 quad), electric clothes dryers (0.58 quad), and dishwashers (0.28 quad). Does not include water heating energy.

[j] Includes residential small electric devices, heating elements, motors, swimming pool heaters, hot tub heaters, outdoor grills, and natural gas outdoor lighting. Includes commercial service station equipment, ATMs, telecommunications equipment, medical equipment, pumps, emergency electric generators, combined heat and power in commercial buildings, and manufacturing performed in commercial buildings.

[k] Energy adjustment EIA uses to relieve discrepancies between data sources. Energy attributable to the residential and commercial buildings sector, but not directly to specific end uses.

Source: EIA, 2010

189

6.4.2   Materials Use for Building Construction and Maintenance

In addition to requiring substantial amounts of energy for their operation, construction of new residential and commercial buildings accounts for a significant fraction of the non-energy-related materials that are produced, imported, and consumed each year. For example, approximately 3 billion tons of natural aggregates (primarily crushed stone, sand, and gravel) are used each year in the United States (Figure 1-5). In 2006, approximately 60% of natural aggregates were used in road and highway construction and 40% in construction of buildings (Sullivan, 2006), indicating that 1.2 billion tons of aggregates are used for building construction in the United States (4 tons per person per year).

Metals are another important category of materials use in buildings. According to the American Iron and Steel Institute (USGS, 2011b), 21% to 24% (depending on year) of iron and steel is shipped to the construction industry, and approximately 60% of the value of new construction put in place in 2011 is for residential and nonresidential buildings (USDC, 2011). Thus, about 15% of annual U.S. iron and steel consumption is for building construction. Iron and steel, in turn, account for 95% of metals production in the United States and in the world (USGS, 2011a). Both primary (new extraction) and recycled sources of metal are used, with roughly 50% of metals production from each (USGS, 2009). If the degree to which metals are recycled is to be substantially increased, methods for the reuse of building materials will be needed.

Building construction and use also drive other types of materials use. Water use in buildings consumes approximately 13% of potable water supplies, mostly for residential as opposed to commercial applications (U.S. EPA, 2009; USGS, 1995). Approximately 60% of forest products, including lumber, plywood/veneer, pulp products, and fuel wood, are used in building construction and operation (U.S. Forest Service, 2003). Table 6-7 summarizes the importance of buildings to the consumption of materials. Overall, it is clear that buildings represent one of the most important targets for sustainable engineering design.

Question 17

Using the data in Table 6.7, estimate the per capita annual and daily materials use for buildings in the United States, assuming that the population of the United States is 300 million.

Table 6-7   Contribution of Buildings to Annual Consumption of Materials in the United States

Material	Total Use (10^6 Metric Tons)	Percent Used in Buildings
Construction materials (natural aggregates)	2910	40%
Metals (iron and steel)	153	15%
Forest products	184	60%

Source: USGS, 2009

6.4.3 Design of Buildings for Sustainability

In order to design buildings for sustainability, additional considerations beyond energy consumption, CO_2 emissions, and materials use must be included. Issues such as indoor air quality, water use efficiency, worker productivity, and even factors such as biodiversity and control of urban warming by buildings become important design considerations. In addition, beyond the direct interaction of people with residential and commercial buildings, there is a complex interplay between buildings and societal structure. For example, the presence or absence of a public transportation infrastructure can affect decisions on commercial or residential building location as well as decisions on the use of public transportation by the building's occupants.

The remainder of this case study will explore selected aspects of sustainable building design. To guide this exploration, design criteria from the U.S. Green Building Council's (USGBC) Leadership in Energy and Environmental Design (LEED) certification program will be used (USGBC, 2011). The LEED program is an internationally recognized building certification program for architects, contractors, owners, and operators. It provides a framework for building design, construction, operation, and maintenance to achieve reductions in environmental impacts and improvement in conditions for building occupants.

Leadership in Energy and Environmental Design (LEED)

The LEED rating system is meant for new and existing commercial, institutional, and residential buildings, and also for entire neighborhood developments. There are currently 8000 LEED-qualified buildings in the Unites States (USGBC, 2011a). After the release of LEED Green Building Rating System Version 1.0 in 1998, several updates to the rating system have occurred, in 2000, 2002, and finally with the most recent Version 2.2 in 2005 (USGBC, 2011b). There are separate rating systems for different building types and project scopes, including

- LEED for Existing Buildings: Operation and Maintenance
- LEED for New Construction
- LEED for Core and Shell
- LEED for Schools
- LEED for Neighborhood Development
- LEED for Retail
- LEED for Health Care
- LEED for Homes
- LEED for Commercial Interiors

Each rating system listed includes five environmental categories: Sustainable Sites, Water Efficiency, Energy and Atmosphere, Material Resources, and Indoor

Environmental Quality. Two additional nonenvironmental human effects categories are also included in the rating system: Innovation in Design and Regional Priority.

In the LEED rating system, each category is subdivided into "credit" subcategories, and one or more points are assigned to each credit subcategory. Total possible points in each ranking system for the five environmental categories is 100 base points; however, 10 additional points can be awarded for the Innovation in Design and Regional Priority categories. The allocation of points among the credit subcategories is based on an analysis of potential environmental impacts and human benefits relative to a set of impact indicators. These impact indicators include GHG emissions, fossil fuel use, toxins and carcinogens, air and water pollutants, and indoor environment. The assessment of potential environmental impacts uses the TRACI impact assessment model from the U.S. EPA (see Chapter 5, Section 5.4) and then applies additional weightings to each subcategory credit from the National Institute of Standards and Technology (NIST), similar to the way that weightings are applied in life-cycle assessment (see Chapter 5 and the "valuation" discussion). The decision on how many points to assign to each credit subcategory is based on a consensus process with a committee of sustainability experts.

The USGBC LEED program certifies that building projects achieve certain levels of desired outcomes according to a numerical scale. For example, in retail applications, the following point scale applies to both design and construction phases:

Certified	40–49 points
Silver	50–59 points
Gold	60–79 points
Platinum	80 points and above

The distribution of these points among the major building categories is dependent on application, but insights into the workings of the LEED rating system can be gained by closer inspection of one application: Retail buildings (LEED, 2010). Here the points are distributed as follows:

Sustainable Sites	26 points
Water Efficiency	10 points
Energy and Atmosphere	35 points
Material Resources	14 points
Indoor Environmental Quality	15 points
Total	100 points

The greatest number of points, and therefore the greatest emphasis in green building design and operation, relates to energy consumption/atmosphere protection and sustainable sites. The following examples will highlight the energy savings opportunities from implementing some of the actions in the LEED design and operation program.

High-Efficiency Lighting

One of the credit subcategories within the Energy and Atmosphere category for LEED Retail buildings is Optimize Energy Performance, where up to 19 points may be awarded. Table 6-6 indicates that lighting is the second most important energy use category in U.S. buildings, contributing 13.4% to the annual building energy consumption total of 40.40 quadrillion BTU/yr, or 5.7% of the annual U.S. energy consumption of 94.6 quadrillion BTU/yr.

Stackhouse and Fan (2009) investigated the replacement of T12 fluorescent bulbs and fixtures with the more efficient T8 version. Existing T12 systems are generally 34-watt lamps driven by an energy-efficient magnetic ballast. A two-lamp one-ballast T12 fixture typically operates at 72 watts (34 watts for each lamp and 4 watts for the ballast) and puts out 2350 lumens. T8 lamps are thinner than T12 lamps. High-performance T8 lighting consists of high-lumen long-life lamps (2250 lumens) coupled with electronic ballasts. This combination provides nearly the same amount of light as a standard dual 34-watt T12 system but uses only 48 watts.

Functional Unit/System Boundaries. Because both bulbs last 20,000 hours and put out about the same light intensity, a suitable functional unit for comparison would be the light output over the life of a pair of bulbs/fixture units. System boundaries for this analysis are from "cradle to use" but do not include impacts of disposal. Issues can arise in the disposal of fluorescent lamps, but in this analysis it will be assumed that the lamps have comparable disposal issues.

Inventory. The inventory of inputs for each bulb system is shown in Table 6-8. The materials identified in Table 6-8 are also the names of ecoprofiles in the ecoinvent database in the LCA software tool SimaPro 7.2. The input amounts in the second column include mathematical formulas for the distribution of total mass between metals and other materials for ballast, bulb, and fixture, lens, and packaging. Electricity demand over the life of the bulb is included in the inputs table, assuming average U.S. grid electricity.

Impact Assessment. Greenhouse gas emissions were converted to equivalents of CO_2 emissions by using global warming potentials (GWPs) from the IPCC 2007 100a method in SimaPro 7.2, where GWPs for CO_2, CH_4, and N_2O are 1, 25, and 298, respectively. GWPs for refrigerants and certain chlorinated and brominated solvents were also included in the analysis. Energy consumption by primary energy type was included using the Cumulative Energy Demand method in SimaPro 7.2.

LCA Results. Figure 6-9 shows the total emission of greenhouse gases expressed in CO_2 equivalents for the life of the pairs of T12 and T8 bulbs/fixtures. Most of the emissions, over 99%, are due to electricity use during the life of the bulb, ballast, and fixture. Production of the bulb systems, including packaging, is 0.1% to 0.2% of total life-cycle emissions. Savings of GHG emissions for the same functional unit for the T8 compared to the T12 is 33.3% = ((1953.8 − 1302.9)/1953.8) *100. Figure 6-10 shows the total energy demand, expressed in megajoule equivalents, reported by primary energy type for the T12 and T8 bulb systems. Most energy is provided by fossil resources, followed by nuclear, with very small contributions by

Table 6-8 Inputs of Materials and Utilities to a Pair of T12 and T8 units

T12	Materials/Utilities[a]			Comments
	Steel, low-alloyed, at plant/ RER S	275*0.99	g	T12 magnetic ballast weight
	Polyethylene, HDPE, granulate, at plant/RER S	275*0.01	g	
	Glass tube, borosilicate, at plant/DE S	54*2*0.9	g	T12 bulb weight, 2 bulbs per system
	Steel, low-alloyed, at plant/ RER S	54*2*0.1	g	
	Steel, low-alloyed, at plant/ RER S	642*0.98	g	T12 4-bulb fixture weight (no bulbs)
	Polyethylene, HDPE, granulate, at plant/RER S	642*0.02	g	
	Polyethylene, HDPE, granulate, at plant/RER S	210	g	T12 lens weight
	Packaging, corrugated board, mixed fiber, single wall, at plant/RER S	6	g	T12 magnetic ballast box
	Packaging, corrugated board, mixed fiber, single wall, at plant/RER S	90	g	T12 24 8-foot bulb box
	Electricity U.S. mix	144*20.000/1000	kWh	20,000 hrs operation
T8	Materials/Utilities			Comments
	Steel, low-alloyed, at plant/ RER S	47*0.99	g	T-8 magnetic ballast weight
	Polyethylene, HDPE, granulate, at plant/RER S	47*0.01	g	
	Glass tube, borosilicate, at plant/DE S	14*4*0.9	g	T-8 bulb weight, 4 bulbs per system
	Steel, low-alloyed, at plant/ RER S	14*4*0.1	g	
	Steel, low-alloyed, at plant/ RER S	762*0.98	g	T8 4-bulb fixture weight (no bulbs)
	Polyethylene, HDPE, granulate, at plant/RER S	762*0.02	g	
	Polyethylene, HDPE, granulate, at plant/RER S	196	g	T8 lens weight
	Packaging, corrugated board, mixed fiber, single wall, at plant/RER S	6	g	T8 magnetic ballast box
	Packaging, corrugated board, mixed fiber, single wall, at plant/RER S	46	g	T8 24 8-foot bulb box
	Electricity U.S. mix	96*20.000/1000	kWh	20,000 hrs operation

[a] These are names of inventory profiles from the ecoinvent database (ecoinvent, 2011).

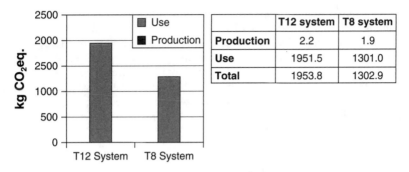

	T12 system	T8 system
Production	2.2	1.9
Use	1951.5	1301.0
Total	1953.8	1302.9

Figure 6-9   Greenhouse gas emissions expressed in CO_2 equivalents for T12 and T8 bulbs

renewable energy sources. Savings of energy for the same functional unit for the T8 compared to the T12 is also 33.3% = ((32,035.7 − 21,365.8)/32,035.7) *100.

LCA Interpretation. Significant savings of energy resources and GHG emissions can be achieved through innovation in lighting systems for buildings. Other types of high-efficiency lighting besides the fluorescent systems discussed here include light-emitting diodes (LEDs), which achieve even higher energy efficiencies than fluorescent bulbs. Also, the effective use of natural light often can enhance artificial lighting systems, such as the use of ceiling skylights in many big box stores (look up the next time you are in one).

Question 18

Assume that the savings in energy consumption calculated in this case study are representative of the magnitude of savings to be expected from lighting improvements in buildings. Estimate the annual reductions of energy demand that could be achieved

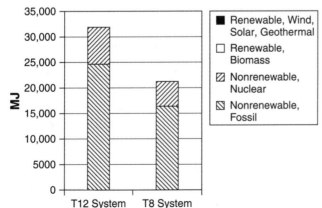

Figure 6-10   Cumulative energy demand for T12 and T8 bulb life cycles

in the United States by improving building efficiency. Use of even more efficient lighting technologies such as LED and natural lighting will reduce lighting energy demands even more. Perform a literature search for estimates of energy efficiencies of LED lighting systems and estimate the energy savings possible from using them, as compared to fluorescent lighting systems. Note that purchase costs of advanced lighting systems such as LEDs can be much higher than conventional alternatives. Estimate the time that would be required to recoup the purchase costs of LED systems, based on savings in energy costs associated with changing from fluorescent to LED lighting.

Simulating Energy Use in Buildings

Besides using high-efficiency lighting in buildings, another approach to improving building energy performance is to use energy balance simulations to optimize the effects of thermal insulation, air exchange rates, and other building design elements and operating procedures. These types of analyses can identify opportunities for improvements in the energy-related impacts of building operation.

Energy Balance Simulation. Energy balance models for buildings are valuable to engineers and architects because these models allow for a better understanding of the major causes of energy demand and can point to design decisions that reduce building energy consumption. There exist a large number of software tools for modeling building energy demand and environmental/economic impacts, and a summary of tools is available at a U.S. DOE Buildings Energy Software Tools Directory Web site (U.S. DOE, 2011a). Some of these software tools are free to download and use, others are to be used online, and others still are commercial products. One such online tool is hosted at the University of Texas, Austin, in the College of Engineering and is called the Building Mass and Energy Balance (BMEB) model (www.ce.utexas.edu/bmeb/). This model can be used to predict the change in energy use rate for residential buildings for changes in window design, ceiling and insulation, air exchange rate, and internal sources of heat from cooking. The model assumes a building location in either Austin or Chicago; the Chicago structure includes a basement that doubles the residence volume compared to the Austin structure.

The model includes conductive heat exchange through ceiling, walls, and floor between the building interior and outside environment through a series of thermal resistances caused by materials, and also in parallel through wall support structures (studs, for example). The conduction component of the model is set up to calculate the maximum heating and cooling loads that a structure must be designed to accommodate, and so the temperature differences in the heat load calculation use the maximum and minimum outside temperatures as opposed to the annual average value. Both sensible and latent heat (differences in air humidity) effects are taken into account. The effects of outside and inside wall convective resistances are also taken into account in the conduction component. Air exchange between the inside and outside of the residence is taken into account, factoring in the heat capacity of the air, rate of air exchange, and temperature difference. Internal sources of heat that can modify building energy loads from people and cooking activities are also taken into account in the model. A detailed description of all energy balance mechanisms, including equations, parameters, and input data, is available on the Web site and should be reviewed prior to using the model. Figure 6-11 shows the model interface and input fields for choices to be made in the model. Results for heating and cooling loads are also displayed, reflecting the choices made to model inputs in the parameter fields.

The effects of changing the air exchange rate on building energy loads are displayed in Table 6-9 for the Austin location. Building cooling loads are larger than heating loads for this location, and changes in these loads are strongly affected by air changes per hour (ACH, the air flow into/out of the building divided by the volume of the building). ACH increases cooling loads more strongly than heating loads, mostly because of the large effect of latent heat, which in this model applies only to cooling loads, as described on the model Web site.

Question 19

Extrapolating the results from Table 6-9, estimate the total load when ACH equals zero. Do you expect the relationship between total load and ACH to be linear? Why or why not? At what ACH level does the energy demand due to air exchange become more than 50% of the total load?

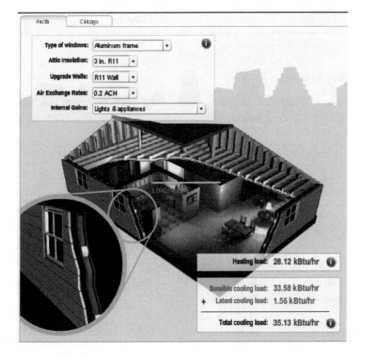

Figure 6-11 Building Mass and Energy Balance model interface (Corsi, 2011)

Question 20

Use the BMEB model to determine which of the available building material choices or operational choices affect energy load the most: type of window, attic and wall insulation, air exchange, or internal energy gains. Generate tables similar to Table 6-9 for each of these choices.

Question 21

Air quality in buildings is often improved by increasing the air exchange rate so that air pollutants generated inside the building can be exchanged with cleaner outside air. Based on the results obtained from the BMEB model, comment on the potential trade-offs between indoor air quality and potential building energy reduction strategies.

Question 22

The energy load results from the BMEB model are most relevant to small residential buildings. How will these results translate to larger residential or commercial high-rise buildings? Consider the wall and ceiling/ground floor area versus the air volume inside the building in your answer. Specifically, by what factor does the area for heat transfer increase in going from a one-story to a ten-story building compared to the factor of increase in the inside air volume? Assume that each story increases the building height by 4 meters, and that each wall in the length and width directions of the structure is 30 meters. Does relative importance of ACH increase, decrease, or stay the same when structure height increases?

Impacts of Building Materials

As this case study has illustrated, building energy use and the materials used in constructing buildings can have significant environmental footprints. Materials are also

Table 6-9   Energy Load Changes from Changes in Air Exchange in Austin, TX
(kBTU = 10^3 BTU)

	Air Changes per Hour (ACH)		
	0.2	0.5	1.3
Heating load (kBTU/hr)	28.12	31.56	40.71
Cooling load (kBTU/hr)	35.13	40.62	55.24
Sensible cooling load	33.58	37.01	46.16
Latent cooling load	1.56	3.61	9.07
Total load (kBTU/hr)	63.25	72.18	95.95

used in the routine operation of buildings, and this material use can have impacts as well. Refrigerants are an example of these types of impacts. Refrigerants used in building cooling can be potent greenhouse gases, and their leakage can be a factor in the overall environmental performance of a building.

The LEED framework allows credit points for selecting environmentally preferable refrigerants for air conditioning, refrigeration, and ventilation that eliminate emission of compounds that contribute to ozone depletion and climate change. The base building equipment must comply with the following formula which sets maximum thresholds for combined ozone depletion and global warming potentials (USGBC, 2009):

$$[LCGWP + (LCODP \times 10^5)] \leq 100$$

LCGWP: Life-cycle direct global warming potential (lb CO_2/ton-year)
LCODP: Life-cycle ozone depletion potential (lb CFC 11/ton-year)

$$LCGWP = [GWPr \times (Lr \times Life + Mr) \times Rc]/Life$$
$$LCODP = [ODPr \times (Lr \times Life + Mr) \times Rc]/Life$$

GWPr: Global warming potential of refrigerant (0 to 12,000 lb CO_2/lbr)

ODPr: Ozone depletion potential of refrigerant (0 to 0.2 lb CFC 11/lbr)

Lr: Refrigerant leakage rate (0.5% to 2.0%; default of 2% unless otherwise demonstrated)

Mr: End-of-life refrigerant loss (2% to 10%; default of 10% unless otherwise demonstrated)

Rc: Refrigerant charge (0.5 to 5.0 lbs of refrigerant per ton of gross ARI-rated cooling capacity)

Life: Equipment life (10 years; default based on equipment type)

Question 23

Use the formulas for LCGWP and LCODP to find a feasible working fluid for building air conditioning and refrigeration. Use data from the U.S. EPA ozone science Web site for a list of zero ODP substitutes for ozone-depleting substance (ODS) refrigerants (www.epa.gov/ozone/geninfo/gwps.html), and use an average of GWP data in your calculations.

Sustainable Buildings Sites

Up to 10 points out of 100 possible points for LEED Retail buildings is allowed for encouraging alternative transportation by building occupants. This touches on sustainability by potentially affecting personal behaviors, shifting from use of automobiles to higher rates of use of public transportation. Whereas most LEED credits involve the direct effects of building construction and operation on environmental impacts, this component of the LEED program emphasizes the importance of a building's indirect effects on energy and the environment. Points are allotted for proximity of the building's main entrance to commuter rail, light rail, or subway stations, to two or more bus stops, for providing for bicycle parking and for rider change/shower rooms, for preferred parking for low-emitting and fuel-efficient vehicles, for public transportation and bicycle subsidies, and for transportation education programs. The importance of transportation for building occupants can be estimated from national statistics for household passenger vehicle use and the knowledge that most people work in a building of some sort. In 2006, passenger cars consumed 75 billion gallons of fuel out of a total consumption for all highway vehicles of 175 billion gallons (U.S. DOT, 2011). In a recent household transportation survey, approximately 25% of household vehicle miles traveled were to and from work (U.S. DOT, 2009).

Question 24

Assuming that most passenger vehicles are in household use (neglecting taxis, rental cars, etc.), estimate fuel consumption for travel to and from places of work. Compare this to total energy demand and to total direct building energy use (in Table 6-6).

6.4.4 Conclusions on Sustainability of Buildings

This case study highlighted the importance of residential and commercial buildings in energy consumption and materials use on a national scale in the United States. This materials and energy use causes direct and indirect impacts on the environment through emission of pollutants that have global, regional, and local consequences. The design of individual buildings and the manner in which buildings are integrated into other societal infrastructures such as transportation have significant effects on economic, environmental, and societal impacts. In response to the importance of buildings to economic viability and environmental performance on a national scale, a rating system for building construction and operation from the U.S. Green Buildings Council was developed (LEED). This case study explored some of the features of this building rating system with respect to lighting, building energy management, use of materials for refrigeration, and indirect effects of buildings on transportation fuel use by building occupants. Buildings are certainly very important for sustainable engineering because of the influence of people's values and attitudes toward their design and operation, which may span a large range from energy and resource intensive to very low-impact structures. Engineers of all types will continue to develop sustainable technologies and analysis tools for improving the economic, environmental, and societal performance of residential and commercial buildings.

6.5 ADDITIONAL CASE STUDIES

The case studies presented in this chapter are just a few examples of case studies that could be used to illustrate principles from this text. Additional examples can be drawn from the published literature. Some of these case studies are formatted as problems, while others are reports that could serve as the basis for problems. The following list includes examples that have been formatted for use as problems:

- The text *Pollution Prevention: Homework and Design Problems for Engineering Curricula,* published by the American Institute of Chemical Engineers, contains more than 20 case studies related to life cycles, design of materials, thermodynamics, and transport phenomena (Allen et al., 1992).
- A case study of selecting alternative battery materials for electric vehicles is provided by Allen and Steele (1994).
- A case study examining alternative methods of transporting the materials required for a fleece pullover garment, through the manufacturing process, is provided by Hopkins et al. (1994).
- A case study involving a material flow analysis of silver entering San Francisco Bay is provided by Kimbrough et al. (1995).
- A case study of the selection of alternative solvents is provided by Allen (1997a).
- A case study of the development of a system for tracking corporate environmental performance is provided by Allen (1997b).

More case studies, formatted for classroom use, have been developed by participants in faculty workshops sponsored by the Center for Sustainable Engineering (www.csengin.org). The case studies are freely available, without copyright restrictions. Examples include, but are not limited to, estimating the environmental impacts of concrete; water system design, including the use of reclaimed water; methods for incorporating social dimensions of sustainability; landfill power generation; accounting for environmental costs and benefits in a semiconductor facility; assessing environmental product claims; electric power generation; and recycling systems for vehicles.

Finally, additional examples can be drawn from the Green Engineering Web site of the U.S. EPA (www.epa.gov/oppt/greenengineering) and the links available at that site.

REFERENCES

Abraham, M. A., and N. Nguyen. 2003. "Green Engineering: Defining the Principles." Results from the Sandestin Conference. *Environmental Progress* 22(4):233–36.

Allen, D. T. 1997a. "Systematic Design of Substitute Materials: A Solvent Case Study." *Pollution Prevention Review* 7(1):113–18.

————. 1997b. "Measuring Corporate Environmental Performance: The Imperial Chemical Industries Group Environmental Burden System." *Pollution Prevention Review* 7(3):109–14.

Allen, D. T., N. Bakshani, and K. S. Rosselot. 1992. *Pollution Prevention: Homework and Design Problems for Engineering Curricula*. New York: American Institute of Chemical Engineers.

Allen, D. T., and N. Steele. 1994. "P2 Tools for Materials Selection." *Pollution Prevention Review* 4:345–54.

Anastas, Paul T., and J. B. Zimmerman. 2003. "Design through the 12 Principles of Green Engineering." *Environmental Science & Technology* 37(5):94A–l01A.

Bozbas, K. 2008. "Biodiesel as an Alternative Motor Fuel: Production and Policies in the European Union." *Renewable and Sustainable Energy Reviews* 12:542–52.

CONCAWE. 2007. *Well-to-Wheels Analysis of Future Automobile Fuels and Powertrains in a European Context: Well-to-Tank Report Version 2c*. WTT App. 1. March. http://ies.jrc.ec.europa.eu/jec-research-collaboration/downloads-jec.html.

Corsi, R. 2011. Building Mass and Energy Balance Software. www.ce.utexas.edu/bmeb. Accessed September 2011.

Davis, S. C., and S. G. Strong. 1993. *Transportation Energy Data Book: Edition 13, Resource and Environmental Profile Analysis of a Woman's Manufactured Apparel Product: Woman's Knit Polyester Blouse*. Prepared for American Fiber Manufacturers Association. Oak Ridge, TN: Oak Ridge National Laboratory.

ecoinvent. 2011. ecoinvent Data v2.2, Swiss Centre for Life Cycle Inventories, www.ecoinvent.ch. Accessed November 12th, 2011.

EIA (Energy Information Administration). 2010. *Annual Energy Outlook 2011 Early Release*. Summary Reference Case Tables, Tables A2, pp. 3–5, Table A4, pp. 9–10, Table A5, pp. 11–12, and Table A17, pp. 34–35; EIA, *National Energy Modeling System (NEMS) for AEO 2011 Early Release*; and EIA, *Supplement to the AEO 2011 Early Release*, Table 32.

Energy Independence and Security Act of 2007 (EISA). 2007. Available at http://frwebgate.access.gpo.gov/cgi-bin/getdoc.cgi?dbname=110_cong_bills&docid=f:h6enr.txt.pdf. Accessed March 2011.

Franklin Associates, Ltd. 1993. *Resource and Environmental Profile Analysis of a Woman's Manufactured Apparel Product: Woman's Knit Polyester Blouse*. Prepared for American Fiber Manufacturers Association.

Ghosh, P., K. J. Hickey, and S. B. Jaffe. 2006. "Development of a Detailed Gasoline Composition-Based Octane Model." *Industrial & Engineering Chemical Research* 45:337–45.

Greenwell, H. C., L. Laurens, R. Shields, R. Lovitt, and K. Flynn. 2010. "Placing Microalgae on the Biofuels Priority List: A Review of the Technological Challenges." *Journal of the Royal Society Interface* 7:703–26. http://rsif.royalsocietypublishing.org/content/7/46/703.full.

Hill, J. N., E. Nelson, D. Tilman, S. Polaski, and D. Tiffany. 2006. "Environmenral, Economic, and Energetic Costs and Benefits of Biodiesel and Ethanol Biofuels." *Proceedings of the National Academy of Sciences* 103:11206–10.

Hopkins, L., D. T. Allen, and M. Brown. 1994. "Quantifying and Reducing Environmental Impacts Resulting from Transportation of a Manufactured Garment." *Pollution Prevention Review* 4:491–500.

Houghton, J., S. Weatherwax, and J. Ferrell. 2006. *Breaking the Biological Barriers to Cellulosic Ethanol: A Joint Research Agenda*. Washington, DC: U.S. Department of Energy. DOE/SC-0095.

IPCC (Intergovernmental Panel on Climate Change). 2006. *2006 IPCC Guidelines for National Greenhouse Gas Inventories*. Prepared by the National Greenhouse Gas Inventories Programme, edited by H. S. Eggleston, L. Buendia, K. Miwa, T. Ngara, and K. Tanabe. Japan: IGES.

————. 2007. *Climate Change 2007, Summary for Policymakers*. A Report of Working Group 1 of the Intergovernmental Panel on Climate Change. Available at www.ipcc.ch.

Kalnes, T., T. Marker, and D. R. Shonnard. 2007. "Green Diesel: A Second Generation Biofuel." *International Journal of Chemical Reaction Engineering* 5, article A48. www.bepress.com/ijcre/vol5/A48.

Kimbrough, D. E., P. W. Wong, and D. T. Allen. 1995. "Policy Options for Encouraging Silver Recovery." *Pollution Prevention Review* 5(4):97–101.

Koers, K. P., T. N. Kalnes, T. Marker, and D. R. Shonnard. 2009. "Green Diesel: A Technoeconomic and Environmental Life Cycle Comparison to Biodiesel and Syndiesel." *Environmental Progress & Sustainable Energy* 28(1):111–20.

Morris, Sydney. 2011. *Running on Air: A Life Cycle Analysis*. Submitted in partial fulfillment of ENG5510, Michigan Technological University, January 7, 2011.

Murphy C. 2011. Chapter 14 in final report to the U.S. Environmental Protection Agency, *Analysis of Innovative Feedstock Sources and Production Technologies for Renewable Fuels*. Cooperative Agreement number XA-83379501-0. Available at www.utexas.edu/research/cem/projects/epa_report.html. Accessed March 2011.

NOAA (National Oceanic and Atmospheric Administration, Earth Systems Research Laboratory). 2010. *Trends in Atmospheric Carbon Dioxide*. www.esrl.noaa.gov/gmd/ccgg/trends/.

NREL (National Renewable Energy Laboratory). 2011. *U.S. Life Cycle Inventory*. Available at www.nrel.gov/lci/database/default.asp. Accessed March 2011.

NSF. 2008. *Breaking the Chemical and Engineering Barriers to Lignocellulosic Biofuels: Next Generation Hydrocarbon Biorefineries,* edited by George W. Huber, University of Massachusetts, Amherst. Washington DC: National Science Foundation, Chemical, Bioengineering, Environmental, and Transport Systems Division.

Peppas, N. A. *1989. One Hundred Years of Chemical Engineering*. Amsterdam: Kluwer.

RFA (Renewable Fuels Association). 2010a. *Ethanol Industry Outlook*. www.ethanolrfa.org/page/-/objects/pdf/outlook/RFAoutlook2010_fin.pdf?nocdn=1.

————. 2010b. *Ethanol Industry Statistics*. www.ethanolrfa.org/pages/statistics#E.

————. 2010c. *U.S. Advanced and Cellulosic Ethanol Projects under Development and Construction*. www.ethanolrfa.org/pages/cellulosic-ethanol.

Searchinger, T., S. Hamburg, J. Melillo, W. Chameides, P. Havlik, D. Kammen, G. Likens, R. Lubowski, M. Obersteiner, M. Oppenheimer, G. P. Robertson, W. Schlesinger, and G. D. Tilman. 2009. "Fixing a Critical Climate Accounting Error. *Science* 326(23):527–28.

Shapouri, H., P. W. Gallagher, W. Nefstead, R. Schwartz, S. Noe, and R. Conway. 2010. *2008 Energy Balance for the Corn-Ethanol Industry*. Washington, DC: U.S. Department of Agriculture. Agricultural Economic Report 846. June.

Sheehan, J. C., V. Camobreco, J. Duffield, M. Graboski, and H. Shapouri. 1998. *Life Cycle Inventory for Biodiesel and Petroleum Diesel for Use in an Urban Bus.* Washington, DC: U.S. Department of Energy, National Renewable Energy Laboratory. NREL/SR-580-24089 UC Category 1503. May.

Shonnard, D. R., L. Williams, and T. N. Kalnes. 2010. "Camelina-Derived Jet Fuel and Diesel: Sustainable Advanced Biofuels." *Environmental Progress & Sustainable Energy.* In press.

Skone, Timothy J., and Kristin Gerdes. 2008. *Development of Baseline Data and Analysis of Life Cycle Greenhouse Gas Emissions of Petroleum-Based Fuels.* Washington, DC: U.S. Department of Energy, National Energy Technology Laboratory, Office of Systems, Analysis and Planning. November 26. Available at www.netl.doe.gov/energy-analyses/pubs/NETL%20LCA%20Petroleum-Based%20Fuels%20Nov%202008.pdf. Accessed July 2011.

Stackhouse, S., and J. Fan. 2009. *Life Cycle Assessment of T12 and T8 Building Lighting Fixtures and Lamps.* Term LCA project for ENG5510 Sustainable Futures 1. Sustainable Futures Institute, Michigan Technological University.

Steen, Eric J., Yisheng Kang, Gregory Bokinsky, Zhihao Hu, Andreas Schirmer, Amy McClure, Stephen B. del Cardayr, and Jay D. Keasling. 2010. "Microbial Production of Fatty-Acid-Derived Fuels and Chemicals from Plant Biomass." *Nature* 463 (7280):559–63.

Sullivan, D. E. 2006. *Materials in Use in Interstate Highways.* Denver, CO: U.S. Geological Survey. http://pubs.usgs.gov/fs/2006/3127/2006-3127.pdf. Accessed July 10, 2011.

U.S. Climate Change Science Program. 2007. *The First State of the Carbon Cycle Report (SOCCR): The North American Carbon Budget and Implications for the Global Carbon Cycle.* November. Available at http://cdiac.ornl.gov/SOCCR/. Accessed September 2011.

USDC (U.S. Department of Commerce). 2011. "Construction at $753.5 Billion Annual Rate." *U.S. Census Bureau News.* May. CB11-118. www.census.gov/const/C30/release.pdf. Accessed July 10, 2011.

U.S. DOE (U.S. Department of Energy). 2011a. *Buildings Energy Software Tools Directory, Energy Efficiency and Renewable Energy.* http://apps1.eere.energy.gov/buildings/tools_directory/subjects_sub.cfm/pagename_menu=whole_building_analysis.

————. 2011b. *U.S. Billion-Ton Update: Biomass Supply for a Bioenergy and Bioproducts Industry*, R. D. Perlack and B.J. Stokes (Leads). ORNL/TM-2011/224. Oak Ridge, TN: Oak Ridge National Laboratory.

U.S. DOE EIA (U.S. Department of Energy, Energy Information Administration). 2010. *Annual Energy Review 2009.* DOE/EIA-0384(2009). www.eia.gov/totalenergy/data/annual/pdf/aer.pdf.

U.S. DOT (U.S. Department of Transportation). 2011. *Data on Motor Vehicle Fuel Consumption and Travel in the U.S., 1960–2006.* www.infoplease.com/ipa/A0004727.html. Accessed July 31, 2011.

————, Federal Highway Administration. 2009. *National Household Travel Survey.* http://nhts.ornl.gov/.

U.S. EPA (U.S. Environmental Protection Agency). 1987. *The Total Exposure Assessment Methodology (TEAM) Study.* EPA 600/S6-87/002.

———. 2009. *Buildings and Their Impact on the Environment: A Statistical Summary.* Revised April 22, 2009. www.epa.gov/greenbuilding/pubs/gbstats.pdf. Accessed July 21, 2011.

———. 2010. "Regulation of Fuels and Fuel Additives: Changes to Renewable Fuel Standard Program; Final Rule." *Federal Register* 75(58):14669–15320.

U.S. Forest Service. 2003. *U.S. Timber Production, Trade, Consumption, and Price Statistics, 1965–2002.* Research Paper FPL-RP-615.

USGS (U.S. Geological Survey). 1995. *Estimated Water Use in the United States in 1995.* http://water.usgs.gov/watuse/pdf1995/html/.

———. 2009. *Use of Minerals and Materials in the United States from 1900 through 2006.* http://pubs.usgs.gov/fs/2009/3008/pdf/FS2009_3008_v1_1.pdf. Accessed July 10, 2011.

———. 2011a. *Iron and Steel Statistics and Information.* http://minerals.usgs.gov/minerals/pubs/commodity/iron_&_steel/. Accessed July 10, 2011.

———. 2011b. *2009 Minerals Yearbook: Iron and Steel [Advanced Release].* http://minerals.usgs.gov/minerals/pubs/commodity/iron_&_steel/. Accessed July 10, 2011.

———. 2011c. *Earthshots: Satellite Images of Environmental Change.* http://earthshots.usgs.gov/Rondonia/Rondonia. Accessed September 2011.

USGBC (U.S. Green Building Council). 2010. *LEED 2009 for Retail: New Construction and Major Renovations.* Washington, DC: U.S. Green Building Council.

———. 2011a. *LEED Projects & Case Studies Directory.* Washington, DC: U.S. Green Building Council. www.usgbc.org/LEED/Project/CertifiedProjectList.aspx. Accessed July 28, 2011.

———. 2011b. www.usgbc.org/. Accessed July 24, 2011.

Index

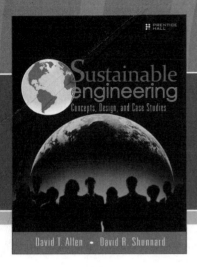

FREE
Online Edition

Safari
Books Online

Your purchase of *Sustainable Engineering* includes access to a free online edition for 45 days through the **Safari Books Online** subscription service. Nearly every Prentice Hall book is available online through Safari Books Online, along with thousands of books and videos from publishers such as Addison-Wesley Professional, Cisco Press, Exam Cram, IBM Press, O'Reilly Media, Que, and Sams.

Safari Books Online is a digital library providing searchable, on-demand access to thousands of technology, digital media, and professional development books and videos from leading publishers. With one monthly or yearly subscription price, you get unlimited access to learning tools and information on topics including mobile app and software development, tips and tricks on using your favorite gadgets, networking, project management, graphic design, and much more.

Activate your FREE Online Edition at
informit.com/safarifree

STEP 1: Enter the coupon code: HLIMMXA.

STEP 2: New Safari users, complete the brief registration form.
Safari subscribers, just log in.

If you have difficulty registering on Safari or accessing the online edition,
please e-mail customer-service@safaribooksonline.com